T0210554

SpringerBriefs in Electrical and Computer
Engineering

More information about this series at http://www.springer.com/series/10059

Miao Pan • Ming Li • Pan Li • Yuguang Fang

Spectrum Trading in Multi-Hop Cognitive Radio Networks

Springer

Miao Pan
University of Houston
Houston, TX, USA

Ming Li
University of Nevada
Reno, NV, USA

Pan Li
Case Western Reserve University
Cleveland, OH, USA

Yuguang Fang
University of Florida
Gainesville, FL, USA

ISSN 2191-8112   ISSN 2191-8120   (electronic)
SpringerBriefs in Electrical and Computer Engineering
ISBN 978-3-319-25629-0   ISBN 978-3-319-25631-3   (eBook)
DOI 10.1007/ 978-3-319-25631-3

Library of Congress Control Number: 2015954942

Springer Cham Heidelberg New York Dordrecht London

Printed on acid-free paper

Springer International Publishing AG Switzerland is part of Springer Science+Business Media (www.springer.com)

To our families

Preface

Cognitive radio (CR) is a revolutionary wireless communication paradigm which releases the spectrum from shackles of authorized licenses and enables secondary users (SUs) to opportunistically access the underutilized licensed spectrum. Due to the great economic value of the spectrum, CR technology has also initiated the spectrum market and promoted a lot of interesting research on spectrum trading designs in cognitive radio networks (CRNs). However, existing spectrum trading approaches mainly focus on per-user-based spectrum trading for single-hop communications and lack deep understanding of multi-hop end-to-end service provision. Correspondingly, there is no such text exclusively on the spectrum trading in multi-hop CRNs today, leaving aspiring researchers and students in this field struggling with limited and scattered literature and sometimes confusing terminology. The goal of this book is to offer some help through accessible presentation of the basic ideas of spectrum trading as well as some related cutting-edge research of spectrum trading in multi-hop CRNs. The target audiences are researchers interested in CR technology and spectrum trading research, in particular graduate students. It is also our hope that this book can be useful to experts as quick reference.

This book starts with an introduction on spectrum trading, state-of-the-art research, and research challenges for spectrum trading in multi-hop CRNs. Then, a novel CRN network architecture tailored for spectrum trading in multi-hop CRNs is introduced. Under this CRN architecture, the transmission opportunity (i.e., a link-band pair)-based spectrum trading is presented, which is beyond the per-user-based spectrum trading in existing literature, and the proof of its economic robustness is provided. Further, this design is extended into session-based spectrum trading under uncertain spectrum supply, and finally an economic robust session-based spectrum trading design is developed and illustrated.

Some of the calculations and proofs involved are mathematical and can be safely skipped in first reading. Nevertheless, we decided to include them because they either illustrate useful analytical skills or provide details that are missing in the original papers. Due to the limited time, space, and of course our knowledge and ability, the content of this book is far from extensive. It only includes closely related literatures that we are mostly familiar with.

We would like to express our greatest appreciation to Dr. Xuemin (Sherman) Shen for providing us the opportunity of writing this short book for Springer. We are also grateful to all our collaborators and colleagues. Finally, we would like to thank Springer, especially Ms. Susan Lagerstrom-Fife, Ms. Jennifer Malat, Ms. Irene Bruce, and Ms. Courtney Clark, for their support in various aspects in the writing and publishing of this book. The book would not come into being without all those efforts and supports.

Houston, TX, USA   Miao Pan
Reno, NV, USA   Ming Li
Cleveland, OH, USA   Pan Li
Gainesville, FL, USA   Yuguang Fang
September 2015

Contents

Acronyms

3-D	Three-dimensional
AWGN	Additive white Gaussian noise
BB	Budget balance
BIP	Binary integer programming
BVR^3	Bidding value-rate requirement ratio
CR	Cognitive radio
CRN	Cognitive radio network
FCC	Federal Communications Commission
IC	Incentive compatibility
IR	Individual rationality
IS	Independent set
LBR	Link-band-radio
LP	Linear programming
MILP	Mixed-integer linear programming
MIS	Maximal independent set
PSP	Primary service provider
PU	Primary user
QoS	Quality of services
SSP	Secondary service provider
SU	Secondary user
TO	Transmission opportunity
TO-AL	Transmission opportunity allocation
TO-SC	Transmission opportunity scheduling
TOST	Transmission opportunity-based spectrum trading
VaR	Value at risk
VBG	Virtual bidder group

Chapter 1
The Network Architecture for Spectrum Trading

Abstract In this chapter, we first describe the motivation for the spectrum trading and briefly summarize the state-of-art research about spectrum trading. Then, we present the research challenges for the spectrum trading in multi-hop cognitive radio networks (CRNs). To address those challenges, we introduce a novel network architecture design for spectrum trading in multi-hop CRNs. The proposed CRN architecture design facilitates the opportunistic spectrum accessing of secondary users (SUs) without cognitive radio (CR) capability, helps to harvest and allocate the available spectrum in efficient way, improves the quality of services (QoS) of multi-hop CR communications, and provides possible approach to guarantee the economic properties of spectrum trading. Under this CRN architecture, we also give content outlines for the rest of the three chapters.

Keywords Network architecture • QoS • End-to-end performance • Economic properties

1.1 Introduction to Spectrum Trading

1.1.1 From Static Spectrum Auction to Dynamic Spectrum Trading

Nowadays, more and more people, families and companies rely on wireless services for their daily life and business, which leads to a booming growth of various wireless networks and a dramatic increase in the demand for radio spectrum. In parallel with that, current static spectrum allocation policy of Federal Communications Commission (FCC) [1, 2, 28] results in the exhaustion of available spectrum, while a lot of licensed spectrum bands are extremely under-utilized. Experimental tests in academia [5, 27] and measurements conducted in industries [26] both show that even in the most crowed region of big cities (e.g., Washington, DC, Chicago, New York City, etc.), many licensed spectrum bands are not used in certain geographical areas and are idle most of the time. Those studies spur the FCC to open up licensed spectrum bands and pursue new innovative technologies to encourage dynamic use of the under-utilized spectrum. As one of the most promising solutions, cognitive

© The Author(s) 2015   1
M. Pan et al., *Spectrum Trading in Multi-Hop Cognitive Radio Networks*,
SpringerBriefs in Electrical and Computer Engineering,
DOI 10.1007/978-3-319-25631-3_1

radio (CR) technology releases the spectrum from shackles of authorized licenses, and enables secondary users (SUs) to opportunistically access to the vacant licensed spectrum bands in either temporal or spatial domain [2, 11, 19, 28, 49].

Due to great economic values of spectrum, the idea of opportunistic using licensed spectrum bands has also initiated the spectrum market and dynamic spectrum trading in cognitive radio networks (CRNs). Dynamic spectrum trading is totally different from the traditional one-time spectrum auction, which is imposed by FCC among big operators (i.e., AT&T, Sprint, T-Mobile, etc.) for long-term static spectrum allocation. It allows primary users (PUs) to sell or lease their vacant spectrum for monetary gains, and SUs to purchase or rent the available licensed spectrum for opportunistic accessing, at anytime, anywhere. Therefore, with joint consideration of uncertain spectrum supply, wireless transmission nature and economic properties, dynamic spectrum trading has promoted a lot of interesting research on the design of spectrum trading systems.

1.1.2   State-of-Art Spectrum Trading Research

Prior work has investigated spectrum trading and harvesting issues from different aspects during last few years. We briefly recapped a few papers here, and some other existing research efforts in this area are referred to in [2, 4, 6, 7, 9, 10, 12–18, 20, 22–24, 29–48, 52–54, 57, 59, 61, 62]. Specifically, in [17], Grandblaise et al. generally describe the potential scenarios and introduce some microeconomics inspired spectrum trading mechanisms, and in [47], Sengupta and Chatterjee propose an economic framework for opportunistic spectrum accessing to guide the design of dynamic spectrum allocation algorithms as well as service pricing mechanisms. From the view of the PUs, Xing et al. in [57], Niyato and Hossain in [29], and Niyato et al. in [30] have well investigated the spectrum pricing issues in the spectrum market, where multiple PUs, whose goal is to maximize the monetary gains with their vacant spectrum, compete with each other to offer spectrum access to the SUs. From the view of the SUs, Pan et al. in [39, 41] have addressed how the SUs optimally distribute their traffic demands over the spectrum bands to reduce the risk for monetary loss, when there is more than one vacant licensed spectrum band. From the view of trading system design, models in game theory, by Wang et al. in [55], Duan et al. in [10], and Zhang and Zhang in [59], and auction designs in microeconomics, by Zhou et al. [60], Zhou and Zheng [62], Jia et al. in [22], Pan et al. in [40], and Wu et al. in [56], are exploited to construct spectrum trading systems with desired properties, such as power efficiency, allocation fairness, incentive compatibility, Pareto efficiency, collusion resistance and so on. Song et al. in [51] and Xu et al. in [58] have studied how to harvest the optimal channel considering the quality of available channels and PUs' activities. Although these designs consider certain features of wireless transmissions (i.e., frequency reuse) and some desired economic properties, they are generally spectrum trading systems for single-hop communications rather than multi-hop communications. They cannot be applied to spectrum trading in multi-hop CRNs due to the very limited concerns on multi-hop CR transmission features, i.e., how to efficiently utilize the purchased spectrum bands in multi-hop CRNs with regard to interference avoidance, flow routing, etc., what kinds of end-to-end quality of services (QoS) can be guaranteed, and so on.

In CR research community, there have been some efforts devoted to cross-layer optimization in multi-hop CRNs as well. Tang et al. in [54] studied the joint spectrum allocation and link scheduling problems with the objectives of maximizing throughput and achieving certain fairness in CRNs. Hou et al. in [21] investigated the joint frequency scheduling and routing problem with the objective of minimizing the network-wide spectrum usage in CRNs. Considering the uncertain spectrum supply, Pan et al. in [43] proposed to model the vacancy of licensed bands as a series of random variables, characterized the multi-hop CRNs with a pair of (α, β) parameters and minimized the usage of licensed spectrum to support CR sessions with rate requirements at certain confidence levels. However, there remains a lack of study to incorporate these multi-hop transmission and uncertain spectrum availability concerns into the designs of spectrum trading.

1.2 Research Challenges for Spectrum Trading

As introduced in previous sections, through spectrum trading, PUs can sell/lease their vacant spectrum for monetary gains, and SUs can purchase/rent the available licensed spectrum if they suffer from the lack of radio resources to support their traffic demands. However, in most existing spectrum trading designs, to trade the licensed spectrum and opportunistically access to these bands, SUs' handsets have to be frequency-agile [28, 46]. It is imperative for the SUs' hand-held devices to have CR capability such as exploring licensed spectrum bands, reconfiguring RF [28, 46], switching frequencies across a wide spectrum range (i.e., from MHz bands such as TV spectrum to GHz bands such as 2 GHz PCS bands or 5 GHz unlicensed bands [8, 46, 50]), and sending and receiving packets over non-contiguous spectrum bands, etc. However, in practice, it may be extremely difficult to embed CR capability into light-weight small-sized radios of SUs' devices. Although some of the desired features may be implemented in small light-weight radios in the future, enormous amount of time and efforts must be spent in hardware designs and signal processing [28, 46, 50]. Besides, even if we can have such CR radios, the prohibitively high price of new fancy CR devices may be discouraging SUs, especially the economically disadvantaged ones, from participating in spectrum trading or even using CRNs. We always expect that certain spectrum trading designs in CRNs can bridge the digital divide and let the low-income people and families enjoy the benefit brought by CR technology without extra cost. Thus, to attract customers for using CR technology to trade spectrum, it is always appreciated to minimize the changes on the handsets of SUs while facilitating spectrum trading to maximize spectral efficiency.

Except for the harsh requirements on SUs' devices, given the harvested spectrum, previous designs mainly focus on per-user spectrum trading for single-hop communications, i.e., each SU bids and uses the purchased spectrum for communications [22, 47, 56, 60, 61]. Those spectrum trading designs have several critical problems. From the SUs'/bidders' side, it is not clear whom a winning bidder communicates with (the receiver is not clearly specified) and what kinds of QoS the winning bidder will experience. From the spectrum trader's side, it is not clear who is the spectrum trader, how the spectrum trader enforces the "opportunistic accessing" rule without affecting primary services [2, 16], how the spectrum trader improves the spectrum utilization via spectrum trading, how the spectrum trader allocates/sells the harvested spectrum to SUs, and collects the revenue when the spectrum supply from PUs/primary service providers (PSPs) is uncertain, etc. From the network's side, since most SUs' traffic tends to be location based and time varying, and may target at Internet services, it is not clear what kinds of communications CRNs can support through spectrum trading. Current spectrum trading designs mostly tend to favor the single-hop traffic services and lack deep understanding of multi-hop end-to-end service provision. A better spectrum trading design with more concern about special features and wireless nature of CRNs supporting multi-hop CR traffic delivery is in need.

1.3 A Novel Network Architecture for Spectrum Trading in Multi-Hop CRNs

To address those challenges in spectrum trading, in this section, we introduce a novel network CRN architecture for spectrum trading in multi-hop CRNs. As shown in Fig. 1.1a, the proposed mesh-like CRN architecture consists of the SSP, a group of SUs, a set of CR mesh routers, and a collection of available licensed spectrum bands[1] with unequal size of bandwidths. The SSP can be either the existing service provider (e.g., AT&T, T-Mobile, Verizon, etc.) lacking of spectrum within this geographical area or an independent wireless service provider willing to provide better or new kinds of services to SUs. The SSP has its own spectrum, i.e., the SSP's basic bands, and is able to collectively harvest the available licensed spectrum. The SSP has also deployed some CR routers at low cost to facilitate the accessing of SUs. SUs are just end-users not subscribed to primary services. No specific requirements are imposed on the SUs' communication devices. They could be either expensive CR devices or economical devices using accessing technologies that the SSP supports with its basic bands (e.g., laptops using Wi-Fi, cell phones using GSM/GPRS, etc.). The CR routers have CR capability and are equipped with multiple CR radios. SUs report their online traffic requests, which include source and destination, rate requirements and corresponding bidding values of the SUs' sessions, to their nearby CR routers. The fixed CR routers collect these requests

---

[1]Taking the least-utilized spectrum bands introduced in [21], for example, we found that the bandwidth between [1240, 1300] MHz (allocated to amateur radio) is 60 MHz, while bandwidth between [1525, 1710] MHz (allocated to mobile satellites, GPS systems, and meteorological applications) is 185 MHz.

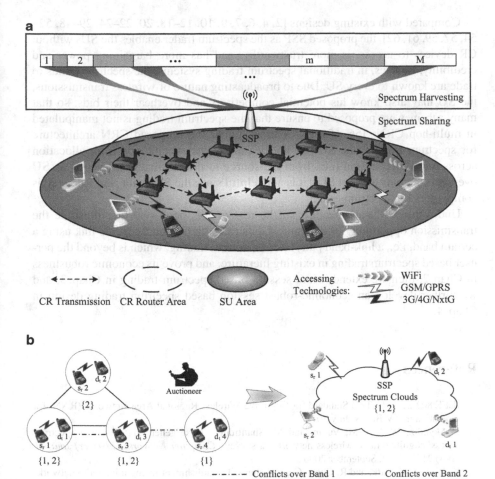

Fig. 1.1 A novel architecture for spectrum trading in multi-hop CRNs. (**a**) Network architecture for spectrum trading. (**b**) A schematic for comparison between per-user based spectrum trading designs and the proposed one

from different end-users and report them to the SSP. If an SU has CR device, it can communicate with a CR router over either basic band or the harvested spectrum band. If not, a CR router will tune to the basic band and communicate with the SU over the basic band. Just because this may potentially reduce the transmission range for the last-hop communications to the end SUs using the basic bands, the frequency reuse for the basic band can be significantly increased with proper frequency planning. The SSP exchanges small-size control messages with CR routers over the common control channel [3, 25]. Depending on the bidding values, rate requirements and the available spectrum resources, the SSP makes decisions on the spectrum trading and harvesting, and jointly conducts scheduling and routing among CR routers for SUs' traffic delivery.

Compared with existing designs [2, 4, 6, 7, 9, 10, 12–18, 20, 22–24, 29–48, 52–54, 57, 59, 61, 62], the proposed SSP as the spectrum trader enables the SUs without CR devices to access, is more trustworthy, and has more bargaining power and credibility. Besides, in traditional spectrum trading systems, the spectrum bands to trade are known to every SU. Due to broadcasting nature of wireless transmissions, the SU may also know his potential competitors and overhear their bids, so that many schemes are proposed to ensure that the spectrum trading is not manipulated in multi-hop CRNs [56, 60]. By contrast, under the proposed CRN architecture for spectrum trading, the SU has no idea about the specific spectrum allocation across the whole session (i.e., from the source to the destination). Even if an SU overhears the bids of other SUs, it is not helpful since the SU is not sure who are his competitors for spectrum usage.

Under this CRN architecture, in the rest of this book, we first illustrate the transmission opportunity (the permit of data transmission on a specific link using a certain band, i.e., a link-band pair) based spectrum trading, which is beyond the per-user based spectrum trading in existing literature, and prove its economic robustness in Chap. 2, then we extend it into session based spectrum trading in Chap. 3, and we further develop an economic-robust session based spectrum trading design in Chap. 4.

References

1. IEEE 802.22-2011(TM) Standard for Cognitive Wireless Regional Area Networks (RAN) for Operation in TV Bands, July 2011.
2. I. Akyildiz, W. Lee, M. Vuran, and M. Shantidev. Next generation/ dynamic spectrum access/ cognitive radio wireless networks: a survey. *Computer Networks (Elsevier) Journal*, 50(4):2127–2159, September 2006.
3. K. Bian, J.-M. Park, and R. Chen. Control channel establishment in cognitive radio networks using channel hopping. *IEEE Journal on Selected Areas in Communications*, 29(4):689–703, April 2011.
4. L. Cao and H. Zheng. Distributed spectrum allocation via local bargaining. In *Proc. of IEEE Communications Society Conference on Sensor and Ad Hoc Communications and Networks, SECON, 2005*, Santa Clara, CA, Sept. 2005.
5. D. Chen, S. Yin, Q. Zhang, M. Liu, and S. Li. Mining spectrum usage data: a large-scale spectrum measurement study. In *Proc. of international conference on Mobile computing and networking, ACM Mobicom, 2009*, Beijing, China, September 2009.
6. R. Chen, J.-M. Park, and J. H. Reed. Defense against primary user emulation attacks in cognitive radio networks. *IEEE Journal on Selected Areas in Communications*, 26(1), January 2008.
7. L. Deek, X. Zhou, K. Almeroth, and H. Zheng. To preempt or not: Tackling bid and time-based cheating in online spectrum auctions. In *Proc. of IEEE Conference on Computer Communications, INFOCOM 2011*, Shanghai, China, April 2011.
8. Defense Advanced Research Projects Agency (DARPA). The neXt generation program (XG) official website.
9. L. Ding, T. Melodia, S. Batalama, and J. Matyjas. Distributed routing, relay selection, and spectrum allocation in cognitive and cooperative ad hoc networks. In *Proc. of IEEE Communications Society Conference on Sensor and Ad Hoc Communications and Networks, SECON, 2010*, Boston, MA, June 2010.

10. L. Duan, J. Huang, and B. Shou. Cognitive mobile virtual network operator: Investment and pricing with supply uncertainty. In *Proc. of IEEE Conference on Computer Communications, INFOCOM 2010*, San Diego, CA, March 2010.
11. FCC. Spectrum policy task force report. Report of Federal Communications Commission, Et docket No. 02-135, November 2002.
12. Z. Feng and Y. Yang. Joint transport, routing and spectrum sharing optimization for wireless networks with frequency-agile radios. In *Proc. of IEEE Conference on Computer Communications, INFOCOM 2009*, Rio de Janeiro, Brazil, April 2009.
13. S. Gandhi, C. Buragohain, L. Cao, H. Zheng, and S. Suri. A general framework for wireless spectrum auctions. In *Proc. of IEEE International Symposium on New Frontiers in Dynamic Spectrum Access Networks, DySPAN 2007*, Dublin, Ireland, April 2007.
14. G. Ganesan and Y. G. Li. Cooperative spectrum sensing in cognitive radio: Part i: two user networks. *IEEE Transactions on Wireless Communications*, 6:2204–2213, June 2007.
15. C. Gao, Y. Shi, Y. T. Hou, H. D. Sherali, and H. Zhou. Multicast communications in multi-hop cognitive radio networks. *to appear in IEEE Journal on Selected Areas in Communications*, 2011.
16. L. Giupponi, R. Agusti, J. Perez-Romero, and O. S. Roig. A novel approach for joint radio resource management based on fuzzy neural methodology. *IEEE Transactions on Vehicular Technology*, 57(3):1789–1805, May 2008.
17. D. Grandblaise, P. Bag, P. Levine, K. Moessner, and M. Pan. Microeconomics inspired mechanisms to manage dynamic spectrum allocation. In *Proc. of IEEE International Symposium on New Frontiers in Dynamic Spectrum Access Networks, DySPAN 2007*, Dublin, Ireland, April 2007.
18. O. Guillermo. *Game Theory (3rd ed.)*. Academic Press, San Diego, 1995.
19. S. Haykin. Cognitive radio: brain-empowered wireless communications. *IEEE Journal on selected areas in communications*, 23(2):201–220, 2005.
20. Y. T. Hou, Y. Shi, and H. D. Sherali. Optimal spectrum sharing for multi-hop software defined radio networks. In *Proc. of IEEE Conference on Computer Communications, INFOCOM 2007*, Anchorage, AL, May 2007.
21. Y. T. Hou, Y. Shi, and H. D. Sherali. Spectrum sharing for multi-hop networking with cognitive radios. *IEEE Journal on Selected Areas in Communications*, 26(1):146–155, January 2008.
22. J. Jia, Q. Zhang, Q. Zhang, and M. Liu. Revenue generation for truthful spectrum auction in dynamic spectrum access. In *Proc. of ACM International Symposium on Mobile Ad Hoc Networking and Computing, ACM MobiHoc, 2009*, New Orleans, LA, May 2009.
23. H. Kim and K. G. Shin. Efficient discovery of spectrum opportunities with mac-layer sensing in cognitive radio networks. *IEEE Transactions on Mobile Computing*, 7(5):533–545, May 2008.
24. W.-Y. Lee and I. F. Akyildiz. Optimal spectrum sensing framework for cognitive radio networks. *IEEE Transactions on Wireless Communications*, 7(10):3845–3857, October 2008.
25. B. F. Lo. A survey of common control channel design in cognitive radio networks. *Physical Communication*, 4(1):26–39, March 2011.
26. M. A. McHenry, P. A. Tenhula, D. McCloskey, D. A. Roberson, and C. S. Hood. Chicago spectrum occupancy measurements and analysis and a long-term studies proposal. In *Proc. of TAPAS 2006*, Boston, MA, August 2006.
27. S. M. Mishra, D. Cabric, C. Chang, D. Willkomm, B. V. Schewick, A. Wolisz, and R. W. Brodersen. A real time cognitive radio testbed for physical and link layer experiments. In *Proc. of IEEE International Symposium on New Frontiers in Dynamic Spectrum Access Networks, DySPAN 2005*, Baltimore, MD, November 2005.
28. J. Mitola. Cognitive radio: An integrated agent architecture for software defined radio. Ph.D. Thesis, Royal Institute of Technology, Sweden, May 2000.
29. D. Niyato and E. Hossain. Competitive pricing for spectrum sharing in cognitive radio networks: Dynamic game, inefficiency of nash equilibrium, and collusion. *IEEE Journal on Selected Areas in Communications*, 26(1):192–202, January 2008.

30. D. Niyato, E. Hossain, and Z. Han. Dynamics of multiple-seller and multiple-buyer spectrum trading in cognitive radio networks: A game theoretic modeling approach. *IEEE Transactions on Mobile Computing*, 8(8):1009–1022, Aug. 2009.

31. M. Pan, F. Chen, X. Yin, and Y. Fang. Fair profit allocation in the spectrum auction using the shapley value. In *Proc. of IEEE Global telecommunications conference, Globecom 2009*, Honolulu, HI, USA, December 2009.

32. M. Pan, J. Chen, R. Liu, and P. Zhang. Dynamic spectrum access and joint radio resource management combing for resource allocation in cooperative networks. In *Proc. of IEEE Wireless Communication and Networking Conference, WCNC '07*, Hong Kong, March 2007.

33. M. Pan and Y. Fang. Bargaining based pairwise cooperative spectrum sensing for cognitive radio networks. In *Proc. of IEEE Military Communications Conference, MILCOM '08*, San Diego, CA, USA, November 2008.

34. M. Pan, R. Huang, and Y. Fang. Cost design for opportunistic multi-hop routing in cognitive radio networks. In *Proc. of IEEE Military Communications Conference, MILCOM '08*, San Diego, CA, USA, November 2008.

35. M. Pan, P. Li, and Y. Fang. Cooperative communication aware link scheduling for cognitive vehicular ad-hoc networks. *IEEE Journal on Selected Areas in Communications*, 30(4):760–768, May 2012.

36. M. Pan, P. Li, Y. Song, Y. Fang, and P. Lin. Spectrum clouds: A session based spectrum trading system for multi-hop cognitive radio networks. In *Proc. of IEEE Conference on Computer Communications, INFOCOM 2012*, Orlando, FL, March 2012.

37. M. Pan, S. Liang, H. Xiong, J. Chen, and G. Liu. A novel bargaining based dynamic spectrum management scheme in reconfigurable systems. In *Proc. of International Conference on Systems and Networks Communications, ICSNC 2006*, Tahiti, French Polynesia, November 2006.

38. M. Pan, Y. Long, H. Yue, Y. Fang, and H. Li. Multicast throughput optimization and fair spectrum sharing in cognitive radio networks. In *Proc. of IEEE Global telecommunications conference, Globecom 2012*, Anaheim, CA, December 2012.

39. M. Pan, Y. Song, P. Li, and Y. Fang. Reward and risk for opportunistic spectrum accessing in cognitive radio networks. In *Proc. of IEEE Global telecommunications conference, Globecom 2009*, Miami, FL, December 2010.

40. M. Pan, J. Sun, and Y. Fang. Purging the back-room dealing: Secure spectrum auction leveraging paillier cryptosystem. *IEEE Journal on Selected Areas in Communications*, 29(4), April 2011.

41. M. Pan, H. Yue, Y. Fang, and H. Li. The x loss: Band-mix selection for opportunistic spectrum accessing with uncertain supply from primary service providers. *IEEE Transactions on Mobile Computing*, 11(12):2133–2144, December 2012.

42. M. Pan, H. Yue, C. Zhang, and Y. Fang. Path selection under budget constraints in multi-hop cognitive radio networks. *IEEE Transactions on Mobile Computing*, 99(PrePrints), 2012.

43. M. Pan, C. Zhang, P. Li, and Y. Fang. Joint routing and scheduling for cognitive radio networks under uncertain spectrum supply. In *Proc. of IEEE Conference on Computer Communications, INFOCOM 2011*, Shanghai, China, April 2011.

44. M. Pan, C. Zhang, P. Li, and Y. Fang. Spectrum harvesting and sharing in multi-hop cognitive radio networks under uncertain spectrum supply. *IEEE Journal on Selected Areas in Communications*, 30(2):369–378, February 2012.

45. M. Pan, X. Zhu, and Y. Fang. Using homomorphic encryption to secure the combinatorial spectrum auction without the trustworthy auctioneer. *Wireless Networks*, 18(2):113–128, February 2012.

46. J. H. Reed. *Software Radio: A Modern Approach to Radio Engineering*. Prentice Hall, New York, May 2002.

47. S. Sengupta and M. Chatterjee. An economic framework for dynamic spectrum access and service pricing. *IEEE/ACM Transactions on Networking*, 17(4):1200–1213, Aug. 2009.

48. Y. Shi and Y. T. Hou. A distributed optimization algorithm for multi-hop cognitive radio networks. In *Proc. of IEEE Conference on Computer Communications, INFOCOM 2008*, Phoenix, AZ, April 2008.

49. K. Shin, H. Kim, A. Min, and A. Kumar. Cognitive radios for dynamic spectrum access: from concept to reality. *Wireless Communications, IEEE*, 17(6):64–74, 2010.

50. H. So, A. Tkachenko, and R. W. Brodersen. A unified hardware/software runtime environment for FPGA based reconfigurable computers using borph. In *Proc. of International Conference on Hardware-Software Codesign and System Synthesis*, Seoul, Korea, October 2006.

51. Y. Song, Y. Fang, and Y. Zhang. Stochastic channel selection in cognitive radio networks. In *Proc. of IEEE Global telecommunications conference, Globecom 2007*, Washington, DC, November 2007.

52. Y. Song, C. Zhang, and Y. Fang. Stochastic traffic engineering in multi-hop cognitive wireless mesh networks. *IEEE Transactions on Mobile Computing*, 9(3):305–316, March 2010.

53. J. Tang, S. Misra, and G. Xue. Spectrum allocation and scheduling in dynamic spectrum access wireless networks. In *Proc. of the International Conference on Quality of Service in Heterogeneous Wired/Wireless Networks, QShine 2007*, Vancouver, BC, Canada, August 2007.

54. J. Tang, S. Misra, and G. Xue. Joint spectrum allocation and scheduling for fair spectrum sharing in cognitive radio wireless networks. *Computer Networks (Elsevier) Journal*, 52(11):2148–2158, August 2008.

55. B. Wang, Z. Han, and K. J. R. Liu. Distributed relay selection and power control for multiuser cooperative communication networks using buyer/seller game. In *Proc. of IEEE Conference on Computer Communications, INFOCOM 2007*, Anchorage, AK, May 2007.

56. Y. Wu, B. Wang, K. J. Liu, and T. Clancy. A multi-winner cognitive spectrum auction framework with collusion-resistant mechanisms. In *Proc. of IEEE International Symposium on New Frontiers in Dynamic Spectrum Access Networks, DySPAN '08*, Chicago, IL, October 2008.

57. Y. Xing, R. Chandramouli, and C. Cordeiro. Price dynamics in competitive agile spectrum access markets. *IEEE Journal on Selected Areas in Communications*, 25(3):613–621, April 2007.

58. D. Xu, E. Jung, and X. Liu. Optimal bandwidth selection in multi-channel cognitive radio networks: How much is too much? In *Proc. of IEEE International Symposium on New Frontiers in Dynamic Spectrum Access Networks, DySPAN 2008*, Chicago, IL, October 2008.

59. J. Zhang and Q. Zhang. Stackelberg game for utility-based cooperative cognitive radio networks. In *Proc. of ACM International Symposium on Mobile Ad Hoc Networking and Computing, ACM MobiHoc, 2009*, New Orleans, LA, May 2009.

60. X. Zhou, S. Gandhi, S. Suri, and H. Zheng. ebay in the sky: strategy-proof wireless spectrum auctions. In *Proc. of Mobile Computing and Networking, Mobicom '08*, San Francisco, CA, September 2008.

61. X. Zhou and H. Zheng. Trust: A general framework for truthful double spectrum auctions. In *Proc. of INFOCOM 2009*, Rio de Janeiro, Brazil, April 2009.

62. X. Zhou and H. Zheng. Breaking bidder collusion in large-scale spectrum auctions. In *Proc. of ACM International Symposium on Mobile Ad Hoc Networking and Computing, ACM MobiHoc, 2010*, Chicago, IL, September 2010.

Chapter 2
Economic-Robust Transmission Opportunity Based Spectrum Trading

Abstract Under the network architecture illustrated in Chap. 1, in this chapter, we further introduce a transmission opportunity based spectrum trading scheme, called TOST, which can support multi-hop data traffic, ensure economic-robustness (i.e., incentive compatibility, individual rationality, and budget balance), and generate high revenue for the spectrum trader. Specifically, in TOST, instead of spectrum bands as in traditional spectrum trading schemes, users bid for transmission opportunities (TOs). A TO is defined as the permit of data transmission on a specific link using a certain band, i.e., a link-band pair. The TOST scheme is composed of three procedures: TO allocation, TO scheduling, and pricing, which are performed sequentially and iteratively until the aforementioned goals are reached. We prove that TOST is economic-robust, and conduct extensive simulations to show its effectiveness and efficiency.

Keywords Transmission opportunity • Multi-hop data transmission • Economic-robust

2.1 Problem Formulation

2.1.1 Network Model

Similar to spectrum trading market in existing literature [1, 5, 6, 8–10, 12–15], we consider a spectrum market where a spectrum owner or primary user (PU) acts as a spectrum trader and leases its idle licensed bands $\mathcal{M} = \{1, 2, \ldots, m, \ldots, M\}$ to secondary users (SUs) $\mathcal{N} = \{1, 2, \ldots n, \ldots, N\}$. The SUs are deployed by a secondary service provider (SSP) to fulfill some purposes such as data delivery, data collection, and object tracking. In this chapter, we assume that each SU is equipped with one radio, which means it cannot transmit and receive simultaneously. Suppose there are a set of $\mathcal{L} = \{1, 2, \ldots, l, \ldots L\}$ sessions in the secondary network. We let $s(l)$, $d(l)$, and $r(l)$ denote the source node, destination node, and traffic demand of session $l \in \mathcal{L}$, respectively. $d(l)$ could be multiple hops away from $s(l)$. To deliver the traffics, the SSP asks all the SUs to submit bids to the spectrum trader for transmission opportunities (TOs), each of which is defined as the permit of data

© The Author(s) 2015

M. Pan et al., *Spectrum Trading in Multi-Hop Cognitive Radio Networks*,
SpringerBriefs in Electrical and Computer Engineering,
DOI 10.1007/978-3-319-25631-3_2

transmission on a specific link using a certain band, i.e., a link-band pair. If some SUs win, they pay a price to the spectrum trader and relay data traffic for each other with obtained TOs. The SSP finally pays back all the winning SUs and lets them gain some profits.

Given the network topology, the PU can construct a conflict graph denoted by $G(V, E)$, where V is the vertex set and E is the edge set. In particular, each vertex corresponds to a link-band pair denoted by $((i, j), m)$, where $i \in \mathcal{N}$, $j \in \mathcal{T}_i^m$, and $m \in \mathcal{M}$. Here, $\mathcal{T}_i^m$ is the set of SUs within SU i's transmission range on band m. Besides, two vertices in V are connected with an undirected edge if the corresponding link-band pairs interfere with each other, i.e., if any of the following conditions is true:

- The receiving SU in one link-band pair is within the interference range of the transmitting SU in another link-band pair, given that the both of them are using the same band;
- The two link-band pairs have at least one node in common.

In this conflict graph, an independent set (IS) is a set in which each element is a link-band pair standing for a transmission, and all the elements (or transmissions) can be carried out successfully at the same time. If adding any more link-band pairs into an IS results in a non-independent one, this IS is defined as a maximum independent set (MIS). We denote the set of all the MISs by $\mathcal{I} = \{\mathcal{I}_1, \mathcal{I}_2, \ldots \mathcal{I}_q, \ldots, \mathcal{I}_Q\}$, where $Q = |\mathcal{I}|$, and $\mathcal{I}_q \subseteq V$ for $1 \leq q \leq Q$. We will show later that we do not really need to find all the MISs. We denote the MIS $\mathcal{I}_q$'s time share (out of unit time 1) to be active by λ_q ($\lambda_q \geq 0$). Therefore, if all the data traffics in the network can be supported, we have $\sum_{q=1}^{Q} \lambda_q \leq 1$. Besides, we let $c_{ij}^m(\mathcal{I}_q)$ be the instantaneous transmission rate of the link-band pair $((i, j), m)$ when MIS $\mathcal{I}_q$ is active. Thus, $c_{ij}^m(\mathcal{I}_q)$ is equal to 0 when $((i, j), m) \notin \mathcal{I}_q$, and the capacity of $((i, j), m)$, i.e., c_{ij}^m, otherwise, which will be introduced soon.

We denote SU i's real valuation and bid price for a TO by v_i and c_i, respectively. Here, we consider that SU evaluates different TO the same. In the trading, SUs submit their bids c_i's in a sealed manner, so that no one has access to any information about others' bids. After the spectrum trader receives all the bids, it divides the bidders into different virtual bidder groups (VBGs), each of which is a set of transmitters of all link-band pairs in one MIS. We denote the set of all the VBGs by $\mathcal{G}$. The spectrum trader considers each VBG as a virtual bidder with its group bid being the sum of all SUs' bids in that group, and determines the winning VBGs denoted by $\mathcal{G}_W$. We denote each winning VBG in $\mathcal{G}_W$ by $\mathcal{G}_{W,t}$ ($1 \leq t \leq |\mathcal{G}_W|$), and the set of indexes of the winning VBGs containing SU i by H_i, respectively. We also denote the clearing price for SU i in a winning VBG containing i, say $\mathcal{G}_{W,t}$ ($t \in H_i$), by p_i^t.

2.1.2 Objective of TOST Design

The design of spectrum trading schemes heavily depends on the desired properties. In this chapter, we assume that all SUs are strategic in the sense that they may manipulate their bids to obtain favorable outcomes. We aim to design a spectrum trading scheme that can satisfy three of the most important economic requirements: Incentive Compatibility (IC), Individual Rationality (IR), and Budget Balance (BB), which are defined as follows:

- **Incentive Compatibility (IC)**: The utility function of ST i ($i \in \mathcal{N}$) is a function of all the bids:

$$u_i(c_i, \mathbf{c_{-i}}) = \begin{cases} \sum_{j \in H_i}(v_i - p_i^t), & \text{if } i \text{ wins} \\ & \text{with bid } c_i, \\ 0, & \text{otherwise,} \end{cases} \qquad (2.1)$$

  where $\mathbf{c_{-i}}$ denotes the vector of bids from other STs. Thus, the spectrum trading is IC if for any ST i ($i \in \mathcal{N}$) with any $c_i \neq v_i$ while others' bids are fixed, we have

$$u_i(c_i, \mathbf{c_{-i}}) \leq u_i(v_i, \mathbf{c_{-i}}). \qquad (2.2)$$

- **Individual Rationality (IR)**: The spectrum trading is IR, if no bidder is charged higher than its bid in the trading, i.e., $c_i \leq \sum_{i \in H_i} p_i^t$ for all $i \in \mathcal{N}$.
- **Budget Balanced (BB)**: To make the spectrum trading self-sustained without any external subsidies, the generated revenue of the spectrum trader, i.e., the PU, is required to be non-negative.

We say the spectrum trading is *economic-robust* [3, 15] if it is IC, IR, and BB. Since in this chapter, we consider that the PU leases its own idle spectrum bands without causing quality degradation to its own services, the PU's revenue is the total payment received from the winning SUs, which is always non-negative. Thus, our trading scheme is always BB. We will focus on achieving IC and IR in our spectrum trading scheme design.

2.1.3 Transmission Opportunity's Capacity

Suppose the power spectral density of SU i on band m is a constant and denoted by P_i^m. A widely used model [7] for power propagation gain between SU i and SU j, denoted by g_{ij}, is $g_{i,j} = C \cdot [d(i,j)]^{-\gamma}$, where i and j also denote the positions of SU i and SU j, respectively, $d(i,j)$ refers to the Euclidean distance between i and j, γ is the path loss factor, and C is a constant related to the antenna profiles of the transmitter and the receiver, wavelength, and so on. We assume that the data transmission is successful only if the received power spectral density at the receiver exceeds a threshold P_T^m. Meanwhile, we assume interference becomes non-negligible only if it produces a power spectral density over a threshold of P_I^m at the receiver. Thus, the transmission range of SU i on band m is $R_T^{i,m} = (CP_i^m/P_T^m)^{1/\gamma}$, which comes from $C(R_T^{i,m})^{-\gamma} \cdot P_i^m = P_T^m$. Similarly, based on the interference threshold $P_I^m (P_I^m \leq P_T^m)$, the interference range of SU i is $R_I^{i,m} = (CP_i^m/P_I^m)^{1/\gamma}$, which is no smaller than $R_T^{i,m}$. Thus, different SUs may have different transmission ranges/interference ranges on different channels with different transmission power.

In addition, according to the Shannon–Hartley theorem, if SU i sends data to SU j on link (i,j) using band m, the capacity of the TO, i.e., link-band pair $((i,j),m)$, is

$$c_{ij}^m = W^m \log_2 \left(1 + \frac{g_{ij}P_i^m}{\eta}\right), \tag{2.3}$$

where η is the thermal noise at the receiver. Note that the denominator inside the log function only contains η. This is because of one of our interference constraints, i.e., when node i is transmitting to node j on band m, all the other neighbors of node j within its interference range are prohibited from using this band. We will address the interference constraints in detail in the following section.

2.2 Transmission Opportunity Based Spectrum Trading

In this section, we introduce our proposed transmission opportunity based spectrum trading scheme, called TOST. Recall that in the network there are SUs who need to deliver data traffic to their destinations that are multiple hops away. Thus, the objective of TOST is to choose MISs, and hence VBGs, which can support such traffics and bring high revenue to the spectrum trader. Meanwhile, TOST should be economic-robust. In general, the TOST scheme is composed of three procedures: TO allocation, TO scheduling, and pricing. TO allocation and TO scheduling are performed iteratively until the termination condition is satisfied, which will be discussed in Section 2.2.3. Thereafter, the charging price will be calculated for all winners in the pricing procedure. In what follows, we detail the design of the three procedures, respectively.

2.2.1 Transmission Opportunity Allocation

At the beginning of TO trading, each SU i ($i \in \mathcal{N}$) submits its bid c_i to the spectrum trader. Then, as mentioned before, the spectrum trader can calculate the virtual bid from $\mathcal{G}_q$ as

$$C_q = \sum_{i \in \mathcal{G}_q} c_i. \tag{2.4}$$

The objective of TO allocation is to find out one winning MIS, which corresponds to a winning VBG, that maximizes the virtual bid C_q in each iteration in a monotonic manner. In particular, we will find the VBG with the highest virtual bid in the first iteration, the one with the second highest virtual bid in the second iteration, and so on and so forth until the iteration ends. Such VBGs (MISs) are considered as winning VBGs (MISs) denoted by $\mathscr{G}_W$ ($\mathscr{I}_W$). We will show in Sect. 2.3 that a monotonic TO allocation procedure is critical in achieving the IC and IR properties.

Before formulating the optimization problem, we first list several constraints as follows.

Notice that in the procedure of TO allocation, we do not assume that we know all the MISs, finding which is in fact an NP-complete problem. We denote

$$s_{ij}^m = \begin{cases} 1, & \text{if } i \text{ can transmit to } j \text{ on band } m, \\ 0, & \text{otherwise.} \end{cases}$$

Since an SU is not able to transmit to or receive from multiple SUs on the same frequency band, we have

$$\sum_{j \in \mathscr{T}_i^m} s_{ij}^m \le 1, \text{ and } \sum_{\{i|j \in \mathscr{T}_i^m\}} s_{ij}^m \le 1. \tag{2.5}$$

Besides, an SU cannot use the same frequency band for transmission and reception, due to "self-interference" at physical layer, i.e.,

$$\sum_{\{i|j \in \mathscr{T}_i^m\}} s_{ij}^m + \sum_{q \in \mathscr{T}_j^m} s_{jq}^m \le 1. \tag{2.6}$$

Moreover, recall that in this chapter, we consider each SU is only equipped with a single radio, which means each SU can only transmit or receive on one frequency band at a time. Thus, we can have

$$\sum_{m \in \mathscr{M}} \sum_{\{i|j \in \mathscr{T}_i^m\}} s_{ij}^m + \sum_{m \in \mathscr{M}} \sum_{q \in \mathscr{T}_j^m} s_{jq}^m \le 1. \tag{2.7}$$

Notice that (2.5)–(2.6) will hold whenever (2.7) holds.

In addition to the above constraints at a certain SU, there are also constraints due to potential interference among the SUs. In particular, for a frequency band m, if SU i uses this band for transmitting data to a neighboring SU $j \in \mathscr{T}_i^m$, then any other SUs that can interfere with SU j's reception should not use this band. To model this constraint, we denote by $\mathscr{P}_j^m$ the set of SUs that can interfere with SU j's reception on band m, i.e.,

$$\mathscr{P}_j^m = \{p | d(p, j) \le R_I^{p,m}, p \ne j, \mathscr{T}_p^m \ne \emptyset\}.$$

The physical meaning of $\mathscr{T}_p^m \neq \emptyset$ in the above definition is that SU p has at least one neighbor to which it may transmit data and hence cause interference to SU j's reception. Therefore, we have

$$\sum_{\{i|j \in \mathscr{T}_i^m\}} s_{ij}^m + \sum_{q \in \mathscr{T}_p^m} s_{pq}^m \leq 1 \qquad (\forall p \in \mathscr{P}_j^m). \tag{2.8}$$

Moreover, recall that we need find the tth highest virtual bid in the tth iteration. Thus, in the tth ($t \geq 2$) iteration, we need find the VBG giving the highest virtual bid with the previously found $t-1$ VBGs being excluded. Letting $\mathscr{I}_{W,t}$ and $\mathscr{G}_{W,t}$ denote the MIS and the corresponding VBG that we find in the tth iteration, respectively, we have

$$\sum_{((i,j),m) \in \mathscr{I}_{W,\tau}} s_{ij}^m < |\mathscr{I}_{W,\tau}|, \quad 1 \leq \tau \leq t-1, \tag{2.9}$$

$$\sum_{((i,j),m) \notin \mathscr{I}_{W,\tau}} s_{ij}^m \geq 1, \quad 1 \leq \tau \leq t-1, \tag{2.10}$$

where $|\mathscr{I}_{W,\tau}|$ is the number of elements contained in $\mathscr{I}_{W,\tau}$. Equation (2.9) means that all the link-band pairs in any of the previously found $t-1$ MISs cannot be selected at the same time in the tth iteration, which excludes the previous $t-1$ MISs. Equation (2.10) means that the newly found MIS should contain at least one different link-band pair from any of the previously found $t-1$ MISs.

Consequently, according to the above constraints, the TO allocation (TO-AL) optimization problem finding the VBG with the tth highest virtual bid in the tth iteration can be formulated as follows:

$$\textbf{TO-AL: Maximize} \quad \sum_{i \in \mathscr{N}} \sum_{j \in \mathscr{T}_i} \sum_{m \in \mathscr{M}_i \cap \mathscr{M}_j} s_{ij}^m \cdot c_i$$

$$\textbf{s.t.} \quad \text{Eqs. (2.7)–(2.10)}$$

$$s_{ij}^m = 0 \quad \text{or} \quad 1,$$

where s_{ij}^m's are the optimization variables, c_i are known constants received from the SUs. Note that (2.9) and (2.10) make sure the newly found IS in tth iteration is an MIS and it is different from any MIS found in previous $t-1$ iterations. Besides, (2.9) is in fact always satisfied as long as (2.10) holds. Since s_{ij}^m can only take value of 0 or 1, TO-AL is a binary integer programming (BIP) problem, which can be solved by applying the traditional *branch-and-bound* or *branch-and-cut* [11] approach.

2.2.2 Transmission Opportunity Scheduling

In this chapter, we assume strict allocation [9] in TO trading, i.e., a source node pays the spectrum trader only if its traffic demand is fully satisfied. Thus, the spectrum trader needs to find an optimal way to utilize those winning MISs, trying to deliver all source nodes' traffic by exploring joint scheduling and routing.

Denote the set of the winning MISs found up to the tth iteration by $\mathscr{I}_W^t = \cup_{\tau=1}^t \mathscr{I}_{W,\tau}$. Note that $|\mathscr{I}_W^t| = t$. Letting $f_{ij}(l)$ denote the flow rate of traffic l over link (i,j), where $i \in \mathscr{N}$, $l \in \mathscr{L}$, and $j \in \mathscr{T}_i$ given $\mathscr{T}_i = \cup_{m \in \mathscr{M}} \mathscr{T}_i^m$, the scheduling of the MISs should satisfy the following:

$$\sum_{l \in \mathscr{L}} f_{ij}(l) \le \sum_{q=1}^t \lambda_q \sum_{m \in \mathscr{M}} c_{ij}^m(\mathscr{I}_q). \tag{2.11}$$

We then give routing constraints in the following. Recall that a source SU may need a number of relay nodes to relay its data packets toward the intended destination node. Since routing packets along a single path may not be able to fully take advantage of the local available channels, in this chapter, we employ multi-path routing to deliver packets more effectively and efficiently.

In particular, if SU i is the source of session l, i.e., $i = s(l)$, then we have the following constraints:

$$\sum_{j \ne s(l), s(l) \in \mathscr{T}_j} f_{js(l)}(l) = 0, \tag{2.12}$$

$$\sum_{j \ne s(l), j \in \mathscr{T}_{s(l)}} f_{s(l)j}(l) = r(l). \tag{2.13}$$

The first constraint means that the incoming data rate of session l at its source node is 0. The second constraint means that the traffic for session l may be delivered through multiple nodes on multiple paths, and the total data rates on all outgoing links are equal to the corresponding traffic demand $r(l)$.

If SU i is an intermediate relay node for session l, i.e., $i \ne s(l)$ and $i \ne d(l)$, then

$$\sum_{j \ne s(l), j \in \mathscr{T}_i} f_{ij}(l) = \sum_{p \ne d(l), i \in \mathscr{T}_p} f_{pi}(l), \tag{2.14}$$

which indicates that the total incoming data rates at a relay node are equal to its total outgoing data rates for the same session.

Moreover, if SU i is the destination node of session l, i.e., $i = d(l)$, then we have

$$\sum_{j \ne d(l), j \in \mathscr{T}_{d(l)}} f_{d(l)j}(l) = 0, \tag{2.15}$$

$$\sum_{p \ne d(l), d(l) \in \mathscr{T}_p} f_{pd(l)}(l) = r(l). \tag{2.16}$$

The first constraint means the total outgoing data rate for session l at its destination $d(l)$ is 0, while the second constraint indicates that the total incoming data rate for session l at the destination $d(l)$ is equal to the corresponding traffic demand $r(l)$.

Thus, based on the constraints mentioned above, the TO scheduling (TO-SC) optimization problem in the tth iteration can be formulated as follows:

$$\textbf{TO-SC: Minimize} \quad \sum_{q=1}^{t} \lambda_q$$

$$\textbf{s.t.} \quad \text{Eqs. (2.11)–(2.16)}$$

$$\lambda_q \geq 0 \ (1 \leq q \leq t)$$

$$f_{ij}(l) \geq 0 \ (i \in \mathcal{N}, j \in \mathcal{T}_i, l \in \mathcal{L}).$$

The formulated optimization problem is a linear programming (LP) problem, which can be easily solved by using the simplex method [4]. The optimal result of TO-SC indicates whether the current winning MISs are enough to support the traffic demand. Specifically, if the optimal objective function is no larger than 1, then the traffic can be supported. The solution also shows how to schedule the MISs and route the traffics. Then, the spectrum trader continues to perform pricing as introduced next. Otherwise, it means that the current winning MISs cannot satisfy the traffic demand. Thus, the spectrum trader does not need to perform pricing and another winning MIS is needed from TO-AL.

2.2.3 Pricing

Before we determine the pricing scheme, we would like to discuss the iteration termination condition. As mentioned above, a minimum scheduling length $\sum_{q=1}^{t} \lambda_q$ over all selected VBGs will be obtained via TO-SC in tth iteration. If $\sum_{q=1}^{t} \lambda_q > 1$, it indicates the traffic load cannot be supported by current winning VBGs and more VBG(s) is(are) needed. In this chapter, we set the termination condition as the iteration when $\sum_{q=1}^{t} \lambda_q \leq 1$ is achieved for the first time. As we will show later in the proof of Theorem 2.1, this termination condition plays a critical role in guaranteeing economic-robustness of our scheme.

After fixing the iteration number, i.e., the number of winning VBGs, we set the charging price for each winning SU following the basic idea of VCG auction pricing [2]. Recall that H_i represent the set of indices of winning VBGs containing SU i. The clearing price for i is

$$p_i = \sum_{t \in H_i} p_i^t = \sum_{t \in H_i} \left(C_{t,-\mathbf{i}} - (C_t - c_i) \right), \tag{2.17}$$

where $C_{t,-i}$ stands for the tth highest group bid when SU i is excluded from the tth iteration and $C_t - c_i$ is the sum of bids from the tth highest group bid except c_i. Clearly, the clearing price for winning SU i is irrelevant to its bid. Based on the proposed spectrum allocation and pricing procedure, we are able to prove this spectrum trading framework is economic-robust. We leave its discussion and proof in Sect. 2.3.

2.3 Proof of Economic Properties

In this section, we will demonstrate that our proposed spectrum trading scheme TOST is economic-robust. Since a winning SU can receive multiple spectrums which are determined in multiple iterations of spectrum trading in our scheme, the economic property analysis is more complicated than that in the existing works. To prove the economic-robustness of the proposed spectrum trading scheme, we first have the following lemma:

Lemma 2.1. *When the other SUs' bids, i.e., $\mathbf{c}_{-i}$, are fixed, if a VBG that contains SU i with bid c_i wins in the tth iteration, it also wins by the tth iteration when SU i bids $c_i' > c_i$.*

Proof. Let the winning VBG in the tth iteration containing SU i with bid c_i be $\mathscr{G}_{W,t}(c_i)$. When SU i bids with c_i', denote by t' the iteration in which the same VBG wins. We also denote by t^* the iteration in which TOST procedure terminates. It is possible that $t^* < t$ or $t^* \geq t$. In order to prove this lemma, we discuss under these two scenarios. We first consider the scenario where $t^* \geq t$. For the winning VBG $\mathscr{G}_{W,t}(c_i)$, its group bid is $C_t = c_{-i}^t + c_i$, where $c_{-i}^t = \sum_{j \in \{\mathscr{G}_{W,t} \setminus \{i\}\}} c_j$. When SU i bids $c_i' > c_i$, this VBG's new group bid, denoted by $C_{t'}'$, is $C_{t'}' = C_t - c_i + c_i' > C_t$.

The VBGs losing in all t iterations when i bids with c_i can be divided into two classes, i.e., the VBGs do not contain SU i and the VBGs contain SU i. For any VBG in the first class, we denote its group bids when SU i bids with c_i and with c_i' by C_s and C_s', respectively. Since the other SUs' bids remain the same, we have $C_s' = C_s \leq C_t < C_{t'}'$. Therefore, the VBGs which do not contain i and lose in all t iterations when i bids with c_i will still lose in all t' iteration when i bids c_i'. For any VBG in the second class, it won't beat $C_{t'}'$ as well, since the bids from VBGs containing SU i will increase by the same amount. Therefore, the number of losing VBGs in the t'th iteration when SU i bids c_i' won't be less than that in the tth iteration when SU i bids c_i and thus $t' \leq t$. As $t^* \geq t$, we have $t' \leq t \leq t^*$, i.e., the VBG wins by tth iteration when SU i bids c_i'.

For the second scenario where $t^* < t$, we can prove $t' \leq t$ following the similar approach above. Besides, we need further prove the VBG wins before TOST termination. From the results above, we have $\{\mathscr{G}_{W,k}(c_i') | 1 \leq k \leq t'\} \subset \{\mathscr{G}_{W,k}(c_i) | 1 \leq k \leq t\}$, i.e., all winning VBGs up to t'th iterations when SU i bids c_i' is a subset of all winning VBGs up to tth iterations when SU i bids c_i. Since $\sum_{q=1}^{t-1} \lambda_q > 1$

with the winning VBG set $\{\mathscr{G}_{W,k}(c_i)|1 \leq k \leq t\}$, we have $\sum_{q=1}^{t'-1} \lambda_q > 1$ with the winning VBG set $\{\mathscr{G}_{W,k}(c_i')|1 \leq k \leq t'\}$, and thus $t' < t^*$ according to our iteration termination condition.

In all, we prove the lemma under both scenarios.

Using the above lemma, we can arrive at the following theorem.

Theorem 2.1. *The proposed TOST is incentive compatibility.*

Proof. We need to show that for any SU i with any $c_i \neq v_i$ while the others' bids are fixed, the condition in (2.2) holds. Let $u_i(c_i, \mathbf{c}_{-i})$ and $u_i(v_i, \mathbf{c}_{-i})$ denote SU i's utility when it bids c_i and v_i, respectively. We first consider the scenario where $c_i > v_i$.

- **Case 1**: *SU i loses with both v_i and c_i.* In this case, $u_i(c_i, \mathbf{c}_{-i}) = u_i(v_i, \mathbf{c}_{-i}) = 0$ according to our definition in (2.1). Thus, (2.2) holds.
- **Case 2**: *SU i loses with v_i but wins with c_i.* Obviously, we have $u_i(v_i, \mathbf{c}_{-i}) = 0$. Considering a VBG $\mathscr{G}_{W,t}(c_i)$ containing SU i with bid c_i wins in the tth iteration, we can infer that $\sum_{k=1}^{t-1} \lambda_k > 1$ according to our pricing scheme. Therefore, when SU i is excluded from the tth iteration, the VBG with the group bid of $C_{t,-i}$ wins as well. Since SU i loses when it bids v_i, we have $C_t - c_i + v_i < C_{t,-i}$. Thus, $u_i(c_i, \mathbf{c}_{-i})$ is calculated by

$$u_i(c_i, \mathbf{c}_{-i}) = \sum_{t \in H_i} \left(v_i - C_{t,-i} + (C_t - c_i) \right)$$

$$= \sum_{t \in H_i} \left(((C_t - c_i + v_i) - C_{t,-i} \right) < 0$$

- **Case 3**: *SU i wins with v_i and loses with c_i.* Since $c_i > v_i$, according to Lemma 2.1, this will not happen.
- **Case 4**: *SU i wins with both v_i and c_i.* We denote the set of the indices of the iterations where i wins by bidding c_i and v_i by H_i and H_i', respectively. This case can be further divided into two subcases. In the first subcase, the set of winning VBGs when SU i bids c_i and that when SU i bids v_i, denoted by $\mathscr{G}_W(c_i)$ and $\mathscr{G}_W(v_i)$, respectively, are the same. In the second one, $\mathscr{G}_W(c_i)$ and $\mathscr{G}_W(v_i)$ are different, it means at least one of the winning VBGs when SU i bids v_i loses when i bids c_i according to Lemma 2.1. Since the first subcase can be treated as a special instance for the latter, we focus on the IC proof for the second subcase in the following.

When SU i bids c_i, denote its utility attributed to the common VBGs between $\mathscr{G}_W(c_i)$ and $\mathscr{G}_W(v_i)$ by $u_i^1(c_i, \mathbf{c}_{-i})$ and the utility attributed to the other VBGs by $u_i^2(c_i, \mathbf{c}_{-i})$. When SU i bids v_i, denote its utility attributed to the common VBGs between $\mathscr{G}_W(c_i)$ and $\mathscr{G}_W(v_i)$ by $u_i^1(v_i, \mathbf{c}_{-i})$ which is exactly $u_i(v_i, \mathbf{c}_{-i})$. Then, we have the following results.

First, for those common VBGs between $\mathscr{G}_W(c_i)$ and $\mathscr{G}_W(v_i)$,

$$u_i^1(c_i, \mathbf{c}_{-i}) - u_i^1(v_i, \mathbf{c}_{-i})$$

$$= \sum_{t \in (H_i \cap H_i')} (v_i - C_{t,-i} + C_t - c_i)$$

$$- \sum_{t' \in (H_i \cap H_i')} (v_i - C_{t',-i} + C_{t'} - v_i).$$

Since $\mathscr{G}_W(c_i)$ and $\mathscr{G}_W(v_i)$ are identical, we have $\sum_{t \in (H_i \cap H_i')} v_i = \sum_{t' \in (H_i \cap H_i')} v_i$. In addition, since the bids from any VBG that does not include SU i when i bids either c_i or v_i are the same, the exclusion of SU i from the spectrum trading won't change their relative relationships as well. Therefore, we have $C_{t,-i} = C_{t',-i}$ and thus $\sum_{t \in (H_i \cap H_i')} C_{t,-i} = \sum_{t' \in (H_i \cap H_i')} C_{t',-i}$. In all, we arrive at $u_i^1(c_i, \mathbf{c}_{-i}) = u_i^1(v_i, \mathbf{c}_{-i})$.

Second, for any VBG in $\mathscr{G}_W(c_i)$ but not in $\mathscr{G}_W(v_i)$, since $v_i + C_t - c_i < C_{t,-i}$, we have

$$u_i^2(c_i, \mathbf{c}_{-i}) = \sum_{t \in H_i \setminus (H_i \cap H_i')} (v_i - C_{t,-i} + (C_t - c_i)) < 0.$$

As a result, we can get

$$u_i(c_i, \mathbf{c}_{-i}) - u_i(v_i, \mathbf{c}_{-i})$$

$$= u_i^1(c_i, \mathbf{c}_{-i}) - u_i^1(v_i, \mathbf{c}_{-i}) + u_i^2(c_i, \mathbf{c}_{-i}) < 0.$$

We then consider the scenario where $v_i > c_i$.

- **Case 1**: *SU i loses with both v_i and c_i.* In this case, $u_i(c_i, \mathbf{c}_{-i}) = u_i(v_i, \mathbf{c}_{-i}) = 0$ according to our definition in (2.1). Thus, (2.2) holds.
- **Case 2**: *SU i loses with v_i but wins with c_i.* Since $v_i > c_i$, according to Lemma 2.1, this will not happen.
- **Case 3**: *SU i wins with v_i and loses with c_i.* In this case, we have $u_i(c_i, \mathbf{c}_{-i}) = 0$ and

$$u_i(v_i, \mathbf{c}_{-i}) = \sum_{t \in H_i'} (v_i - C_{t,-i} + (C_t - v_i))$$

$$= \sum_{t \in H_i'} (C_t - C_{t,-i}) > 0.$$

- **Case 4**: *SU i wins with both v_i and c_i.* Similarly, this case can be further divided into two subcases. In the first subcase, $\mathscr{G}_W(c_i)$ and $\mathscr{G}_W(v_i)$ are the same. In the second one, $\mathscr{G}_W(c_i)$ and $\mathscr{G}_W(v_i)$ are different, it means at least one of the winning

VBGs when SU i bids v_i loses when i bids c_i according to Lemma 2.1. Since the first subcase can be treated as a special instance for the latter, we focus on the IC proof for the second subcase in the following.

When SU i bids v_i, denote its utility attributed to the common VBGs between $\mathscr{G}_W(c_i)$ and $\mathscr{G}_W(v_i)$ by $u_i^1(v_i, \mathbf{c}_{-i})$ and the utility attributed to the other VBGs by $u_i^2(v_i, \mathbf{c}_{-i})$. When SU i bids c_i, denote its utility attributed to the common VBGs between $\mathscr{G}_W(c_i)$ and $\mathscr{G}_W(v_i)$ by $u_i^1(c_i, \mathbf{c}_{-i})$ which is exactly $u_i(c_i, \mathbf{c}_{-i})$. Then, we have the following results.

First, for those common VBGs between $\mathscr{G}_W(c_i)$ and $\mathscr{G}_W(v_i)$, we have $u_i^1(c_i, \mathbf{c}_{-i}) = u_i^1(v_i, \mathbf{c}_{-i})$ following the same approach above.

Second, for any VBG in $\mathscr{G}_W(v_i)$ but not in $\mathscr{G}_W(c_i)$, we have

$$u_i^2(v_i, \mathbf{c}_{-i}) = \sum_{t \in H_i' \setminus (H_i' \cap H_i)} (v_i - C_{t,-i} + (C_t - v_i)) > 0.$$

As a result, we can get

$$u_i(c_i, \mathbf{c}_{-i}) - u_i(v_i, \mathbf{c}_{-i})$$
$$= u_i^1(c_i, \mathbf{c}_{-i}) - u_i^1(v_i, \mathbf{c}_{-i}) - u_i^2(v_i, \mathbf{c}_{-i}) < 0.$$

In all, $u_i(c_i, \mathbf{c}_{-i}) \leq u_i(v_i, \mathbf{c}_{-i})$ always hold, and hence the spectrum trading is IC.

Theorem 2.2. *The proposed TOST is individual rationality.*

Proof. For a winning SU i, its utility is expressed as

$$u_i(c_i, \mathbf{c}_{-i}) = \sum_{t \in H_i} (v_i - C_{t,-i} + (C_t - c_i))$$
$$= \sum_{t \in H_i} (-C_{t,-i} + C_t) > 0.$$

For a losing SU i, its utility is 0. In all, SU i always gets non-negative utility.

Theorem 2.3. *The proposed spectrum trading framework is budget balance.*

Proof. For a winning SU i, its clearing price is expressed as

$$p_i = \sum_{t \in H_i} p_i^t = \sum_{t \in H_i} (C_{t,-i} - (C_t - c_i)) \geq 0.$$

For a losing SU i, its clearing price is 0. Thus, the total revenue received by the spectrum trader, i.e., $\sum_{i \in \mathcal{N}} p_i$, is non-negative.

With Theorems 2.1, 2.2 and 2.3, we conclude that the proposed spectrum trading scheme is economic-robust.

2.4 Performance Evaluation

2.4.1 Simulation Setup

In this section, we conduct simulations to evaluate the performance of our proposed spectrum trading scheme TOST. Simulations are carried out in CPLEX 12.4 on a computer with a 2.27 GHz CPU and 24 GB RAM. We randomly deploy SUs in a square network of area 1000 m × 1000 m. There are totally 5 multi-hop sessions in the network, each of which has traffic demand of 1 Mbps. We assume that each bidder's true valuation of (and hence its bid for) unit instantaneous transmission rate is uniformly distributed over $[10^{-6}, 10^{-5}]$. In addition, assume the PU has 3 idle spectrum bands to lease to the SUs, with their bandwidths being 1.0, 1.5 and 2.0 MHz, respectively. Some other important simulation parameters are listed as follows. The path loss exponent is 4 and $C = 62.5$. The noise power spectral density is $\eta = 3.34 \times 10^{-20}$ W/Hz at all nodes. The transmission power spectral density of nodes is $8.1 \times 10^7 \eta$, and the reception threshold and interference threshold are both 8.1η on each spectrum band. Thus, the transmission range and the interference range on each frequency band are both equal to 500 m. Since we have proved that our spectrum trading scheme is economic-robust in the previous section, we demonstrate the *spectrum trading efficiency* and the spectrum trader's revenue in what follows. Note that spectrum trading efficiency is defined as the ratio of the number of finally successfully delivered traffic flows to the total number of traffic flows demanded by the SUs.

2.4.2 Results and Analysis

We first compare the spectrum trading efficiency of the proposed TOST scheme with those of two other trading schemes: one for single-hop data transmission [14], and the other for multi-hop data transmission [16] which greedily assigns spectrum bands to links. We call these two schemes 1-hop spectrum trading and greedy multi-hop spectrum trading, respectively, in our simulations. To make fair comparisons, we compare TOST with these two schemes in single-hop and multi-hop scenarios, respectively.

In the single-hop scenario, each source SU can reach its intended destination SU in one hop, and hence the data traffic can be delivered in one-hop as well. Figure 2.1a gives the results when the number of SUs ranges from 10 to 30 and the number of available spectrums M is equal to 1 and 3. We can find that TOST can achieve much higher spectrum trading efficiency than 1-hop spectrum trading. Particularly, in the case that there is only one available spectrum band, TOST can support two and three traffic flows when the number of SUs is 10 and 15, respectively, while 1-hop spectrum trading cannot support any of the traffic flows. When there are more SUs in the network, TOST can support four traffic flows while 1-hop spectrum trading can

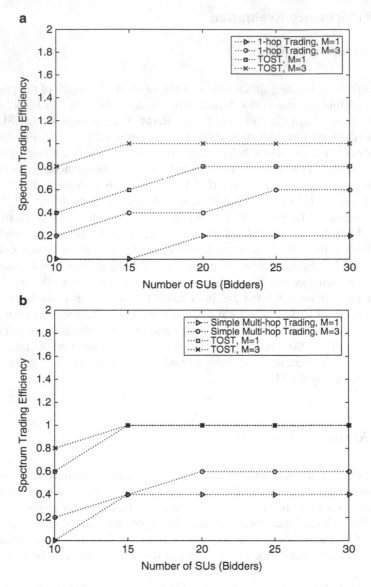

Fig. 2.1 Spectrum trading efficiency comparison with 1-hop trading scheme and greedy multi-hop trading scheme. (**a**) Single-hop data transmission scenario. (**b**) Multi-hop data transmission scenario

only support one of them. In the case that there are three available spectrum bands, TOST can support four flows when there are 10 SUs and all the five flows when there are more SUs, while 1-hop spectrum trading can only support one flow, two flows, and three flows, when there are 10, 15 and 20, and more SUs, respectively.

As we mentioned before, this is because in 1-hop spectrum trading, it is not clear whom a winning SU communicates with and there can be a lot of collisions in the network.

In the multi-hop scenario, each source node needs to deliver data to its destination via multiple hops. The spectrum trading efficiency is shown in Fig. 2.1b when the number of SUs ranges from 10 to 30 and the number of available spectrums M is equal to 1 and 3. In particular, in the case that there is only one available spectrum band, TOST can support three traffic flows when the number of SUs is 10, and all the five traffic flows when there are more SUs in the network. On the other hand, greedy multi-hop spectrum trading cannot support any traffic flows when there are 10 SUs, and only two flows when there are more SUs. Besides, in the case that there are three available spectrum bands, TOST can support four flows when there are 10 SUs and five flows when there are more SUs, while greedy multi-hop spectrum trading can only support one flow, two flows, and three flows, when there are 10, 15, and more SUs, respectively. This is because that we consider transmission opportunities in spectrum trading as well as spectrum scheduling in both frequency and time domains.

References

1. M. Al-Ayyoub and H. Gupta. Truthful spectrum auctions with approximate revenue. In *Proceeding of the IEEE International Conference on Computer Communications (INFOCOM'11)*, Shanghai, China, April 2011.
2. M. Babaiof and N. Nisan. Concurrent auctions across the supply chain. *Journal of Artificial Intelligence Research*, pages 595–629, May 2004.
3. M. Babaioff and W. E. Walsh. Incentive-compatible, budget-balanced, yet highly efficient auctions for supply chain formation. *In Decision Support Systems. In*, pages 64–75, 2004.
4. G. B. Dantzig. *Linear Programming and Extensions*. Princeton University Press, 1963.
5. S. Gandhi, C. Buragohain, L. Cao, H. Zheng, and S. Suri. A general framework for wireless spectrum auctions. In *Proc. of IEEE International Symposium on New Frontiers in Dynamic Spectrum Access Networks, DySPAN 2007*, Dublin, Ireland, April 2007.
6. A. Gopinathan, Z. Li, and C. Wu. Strategyproof auctions for balancing social welfare and fairness in secondary spectrum markets. In *Proceeding of the IEEE International Conference on Computer Communications (INFOCOM'11)*, Shanghai, China, April 2011.
7. Y. T. Hou, Y. Shi, and H. D. Sherali. Spectrum sharing for multi-hop networking with cognitive radios. *IEEE Journal on Selected Areas in Communications*, 26(1):146–155, January 2008.
8. J. Huang, R. A. Berry, and M. L. Honig. Auction-based spectrum sharing. *Journal of Mobile Networks and Applications*, 11(3):405–418, 2006.
9. J. Jia, Q. Zhang, Q. Zhang, and M. Liu. Revenue generation for truthful spectrum auction in dynamic spectrum access. In *Proceeding of ACM MobiHoc*, New Orleans, Louisiana, US, May 2009.
10. C. Kloeck, H. Jaekel, and F. K. Jondral. Dynamic and local combined pricing, allocation and billing system with cognitive radios. In *Proc. of IEEE International Symposium on New Frontiers in Dynamic Spectrum Access Networks, DySPAN 2005*, Baltimore, MD, November 2005.
11. Y. Pochet and L. Wolsey. *Production Planning by Mixed Integer Programming*. Secaucus, 2006.

12. G. Wang, Q. Liu, and J. Wu. Hierarchical attribute-based encryption for fine-grained access control in cloud storage services. In *Proceedings of the 17th ACM conference on Computer and communications security (CCS'10)*, New York, NY, USA, October 2010.
13. W. Wang, B. Li, and B. Liang. District: Embracing local markets in truthful spectrum double auctions. In *Proc. of IEEE Communications Society Conference on Sensor and Ad Hoc Communications and Networks, SECON, 2011*, Salt Lake, UT, June 2011.
14. X. Zhou, S. Gandhi, S. Suri, and H. Zheng. ebay in the sky: Strategy-proof wireless spectrum auctions. In *Proceedings of ACM MobiCom*, San Francisco, CA, USA, September 2008.
15. X. Zhou and H. Zheng. Trust: A general framework for truthful double spectrum auctions. In *Proceeding of the IEEE International Conference on Computer Communications (INFOCOM'09)*, Rio de Janeiro, Brazil, April 2009.
16. Y. Zhu, B. Li, and Z. Li. Truthful spectrum auction design for secondary networks. In *Proceeding of the IEEE International Conference on Computer Communications (INFOCOM'12)*, Orlando, FL, USA, March 2012.

Chapter 3
A Session Based Spectrum Trading System Under Uncertain Spectrum Supply

Abstract Under the same network architecture, in this chapter, we introduce a session based spectrum trading system beyond transmission opportunity based spectrum trading in multi-hop CRNs. As illustrated in Chap. 1, we employ SSP to facilitate the accessing of SUs without CR capability and harvest uncertain spectrum supply. Besides, we also allow the SSP to conduct spectrum trading among CR sessions w.r.t. their conflicts and competitions. Leveraging a three-dimensional (3-D) conflict graph, we mathematically describe the conflicts and competitions among the candidate sessions for spectrum trading. Given the rate requirements and bidding values of candidate trading sessions, we formulate the optimal spectrum trading into the SSP's revenue maximization problem under multiple cross-layer constraints. In view of the NP-hardness of the problem, we develop heuristic algorithms to pursue feasible solutions. Simulation results show the effectiveness and optimality of the proposed algorithms.

Keywords Revenue maximization • Uncertain spectrum availability • Link scheduling • Multi-hop multi-path routing

3.1 Network Model

3.1.1 Network Configuration

We consider a spectrum market consisting of the SSP, a group of SUs, a set of CR mesh routers, and a collection of available licensed spectrum bands[1] with unequal size of bandwidths. Suppose there are $\mathcal{N} = \{1, 2, \cdots, n, \cdots, N\}$ CR mesh routers, each CR mesh router has $\mathcal{H} = \{1, 2, \cdots, h, \cdots, H\}$ radio interfaces, and these CR mesh routers form a set of $\mathcal{L}$ unicast communication sessions according to SUs' requests. Each session has a rate requirement and a corresponding bidding value.

---

[1]Taking the least-utilized spectrum bands introduced in [9], for example, we found that the bandwidth between [1240, 1300] MHz (allocated to amateur radio) is 60 MHz, while bandwidth between [1525, 1710] MHz (allocated to mobile satellites, GPS systems, and meteorological applications) is 185 MHz.

© The Author(s) 2015   27
M. Pan et al., *Spectrum Trading in Multi-Hop Cognitive Radio Networks*,
SpringerBriefs in Electrical and Computer Engineering,
DOI 10.1007/978-3-319-25631-3_3

Denote the source/destination CR router of session $l \in \mathcal{L} = \{1, 2, \cdots, l, \cdots, L\}$ by $s_r(l)/d_t(l)$, and let $(r(l), b(l))$ be the rate requirement-bidding value pair for session $l \in \mathcal{L}$. Assume the SUs' usage of basic bands in the multi-hop CRNs is a priori information. The CR routers are able to use the rest of basic spectrum owned by the SSP. The CR routers are also allowed to communicate with each other by opportunistically accessing to the licensed bands when the primary services are not active, but they must evacuate from these bands when primary services become active.

Considering the geographical location of the CR routers, the available spectrum bands at one CR router may be different from another one in the network. To put it in a mathematical way, let $\mathcal{M} = \{1, 2, \cdots, m \cdots, M\}$ be the band set including the available basic bands and licensed bands with different bandwidths $\mathcal{W} = \{W^1, W^2, \cdots, W^m, \cdots, W^M\}$ for communications, and $\mathcal{M}_i \subseteq \mathcal{M}$ represent the set of available bands at CR router $i \in \mathcal{N}$. $\mathcal{M}_i$ may be different from $\mathcal{M}_j$, where j is not equal to i, and $j \in \mathcal{N}$, i.e., possibly $\mathcal{M}_i \neq \mathcal{M}_j$. Meanwhile, since primary services come back and forth, the spectrum supply from licensed bands is uncertain in the temporal domain. To capture this key feature of spectrum trading in CRNs, let T_{ij}^m denote the available time of band m at CR link (i, j) within one unit time slot, where T_{ij}^m is modeled as a random variable. As shown in [2, 11, 12], the statistical characteristics of T_{ij}^m contain abundant knowledge about band m's spectrum availability at link (i, j) for opportunistic accessing.[2]

3.1.2 Other Related Models in Multi-Hop CRNs

3.1.2.1 Transmission Range and Interference Range

Suppose all CR mesh routers use the same power P for transmission. The power propagation gain [6, 9] is

$$g_{ij} = \gamma \cdot d_{ij}^{-\beta}, \tag{3.1}$$

where β is the path loss factor, γ is an antenna related constant, and d_{ij} is the distance between CR routers i and j. We assume that the data transmission is successful only if the received power at the receiver exceeds the receiver sensitivity, i.e., a threshold P_{Tx}. Meanwhile, we assume interference becomes non-negligible only if it is over a threshold of P_{In} at the receiver. Thus, the transmission range for a CR router is $R_{Tx} = (\gamma P/P_{Tx})^{1/\beta}$, which comes from $\gamma \cdot (R_{Tx})^{-\beta} \cdot P = P_{Tx}$. Similarly, based on

---

[2]Chen et al. in [3] carried out a set of spectrum measurements in the 20 MHz to 3 GHz spectrum bands at four locations concurrently in Guangdong province of China. They used these data sets to conduct a set of detailed analysis on statistics of the collected data, including channel occupancy/vacancy statistics, channel utilization, also spectral and spatial correlation of these measures.

the interference threshold $P_{In}(P_{In} < P_{Tx})$, the interference range for a CR router is $R_{In} = (\gamma P / P_{In})^{1/\beta}$. It is obvious that $R_{In} > R_{Tx}$ since $P_{In} < P_{Tx}$.

In the widely used protocol model [7, 9, 10, 12, 15, 16], the interference range is typically 2 or 3 times of the transmission range, i.e., $\frac{R_{In}}{R_{Tx}} = 2$ or 3. These two ranges may vary with frequency. The conflict relationship between two links over the same frequency band can be determined by the specified interference range. In addition, if the interference range is properly set, the protocol model can be accurately transformed into the physical model as illustrated in [14].

3.1.2.2 Link Capacity/Achievable Data Rate

According to Shannon–Hartley theorem, if CR router i sends data to CR router j on link (i, j) with band m, the capacity of link (i, j) with band m is

$$c_{ij}^m = W^m \log_2 \left(1 + \frac{g_{ij}P}{\eta}\right), \tag{3.2}$$

where η is the ambient Gaussian noise power at CR mesh router j.[3] Depending on different modulation schemes, the achievable data rate is actually determined by the SNR at the receiver and receiver sensitivity [9, 16]. However, in most of existing literature [9, 10, 12], the achievable data rate is approximated by Eq. (3.2), even though this data rate can never be achieved in practice. In this paper, we follow the same approximation. Note that this approximation will not affect the theoretical analysis or performance comparison in this work.

3.2 Optimal Spectrum Trading Under Cross-Layer Constraints in Multi-Hop CRNs

We exploit binary value $\delta(l)$ to denote the success/failure of spectrum trading for session l, i.e.,

$$\delta(l) = \begin{cases} 1, & \text{session } l \text{ is accessed by the SSP}; \\ 0, & \text{session } l \text{ is denied by the SSP}. \end{cases} \tag{3.3}$$

To make the decision of accessing/denying a session $l \in \mathscr{L}$, the SSP must consider both the rate requirement and bidding value of session l. Besides, to effectively utilize the leftover basic spectrum and the harvested licensed spectrum, it is nec-

---

[3]Note that the denominator inside the log function contains only η. This is because of one of our interference constraints, i.e., when CR router i is transmitting to CR router j on band m, then all the other neighbors of router j within its interference range are prohibited from using this band. We will address the interference constraints in detail in the following section.

essary for the SSP to schedule data transmission among different CR mesh routers under joint spectrum assignment, link scheduling and flow routing constraints. In the rest of this section, we first extend the conflict graph [16] to characterize the interference relationship among CR links. Then, based on the extended conflict graph, we mathematically describe link scheduling and flow routing constraints and formulate the spectrum trading into the revenue maximization problem of the SSP under multiple constraints. By relaxing the integral variables, we solve the optimization problem and provide an upper-bound of the SSP's revenue.

3.2.1   Extended Conflict Graph, Cliques, and Independent Sets

3.2.1.1   Construction of Three-Dimensional (3-D) Conflict Graph

Regarding the availability of spectrum bands and radios at CR mesh routers, we introduce a 3-D conflict graph to characterize the interference relationship among CR links in CRNs. Following the definitions in [10], we interpret a CRN as a three-dimensional resource space, with dimensions defined by links, the set of available bands and the set of available radios. In a 3-D conflict graph $\mathscr{G}(\mathscr{V}, \mathscr{E})$, each vertex corresponds to a *link-band-radio* (LBR) tuple, i.e.,

$$\text{link-band-radio:} \quad ((i, j), m, (u, v)),$$

where $i \in \mathscr{N}$, $m \in \mathscr{M}_i \cap \mathscr{M}_j$, $j \in \mathscr{T}_i^m$, $u \in \mathscr{H}_i$, and $v \in \mathscr{H}_j$. Here, $\mathscr{T}_i^m$ is the set of CR mesh routers within CR router i's transmission range. The LBR tuple indicates that the CR router i transmits data to CR router j on band m, where radio interfaces u and v are used at sending CR router and receiving CR router, respectively. Based on the definition of LBR tuples, we can enumerate all combinations of CR mesh routers, the vacant bands and the available radios, which can potentially enable CR communication links.

Different from multi-radio multi-channel networks [10], the availability of bands and radios (i.e., the leftover radios after collecting SUs' traffic) at each CR router in CRNs may be different, i.e., for $i, j \in \mathscr{N}$, maybe $\mathscr{M}_i \neq \mathscr{M}_j$ and $\mathscr{H}_i \neq \mathscr{H}_j$. Similar to the interference conditions in [9, 10, 12], two LBR tuples are defined to interfere with each other if either of the following conditions is true: (1) if two different LBR tuples are using the same band, the receiving CR router of one tuple is in the interference range of the transmitting CR router in the other tuple; (2) two different LBR tuples have the same radios at one or two CR routers.

Note that the first condition not only represents co-band interference but also inherently covers the following two cases: any CR router cannot transmit to multiple routers on the same band; any CR router cannot use the same band for concurrent transmission and reception, due to "self-interference" at the physical layer. Meanwhile, the second condition represents the radio interface conflicts, i.e., a single radio cannot support multiple transmissions (either transmitting or receiving) simultaneously. According to these conditions, we connect two vertices in $\mathscr{V}$ with an undirected edge in $\mathscr{G}(\mathscr{V}, \mathscr{E})$, if their corresponding LBR tuples interfere with each other.

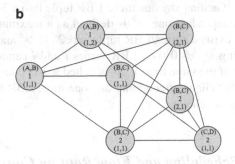

Fig. 3.1 Conflict relationship represented by 3-D conflict graph in CRNs. (**a**) Toy topology in CRNs. (**b**) 3-D conflict graph

For illustrative purposes, we take a simple example to show how to construct a 3-D conflict graph. In this toy CRNs, we assume there are four CR routers with CR transceivers, i.e., A, B, C, and D, and two bands, i.e., band 1 and band 2. Depending on the geographic locations, the set of currently available bands and radios at one CR router may be different from that at another CR router. For example, the currently available band and radio sets for A are $\mathcal{M}_A = \{1\}$ and $\mathcal{H}_A = \{1\}$, and the band and radio sets for B are $\mathcal{M}_B = \{1, 2\}$ and $\mathcal{H}_B = \{1, 2\}$. Furthermore, we use $d(\cdot)$ to represent Euclidean distance and suppose that $d(A, B) = d(B, C) = d(C, D) = d(D, E) = R_{Tx} = 0.5R_{In}$. Given the above assumptions, we can establish the corresponding 3-D conflict graph as depicted in Fig. 3.1b. Here, each vertex corresponds to an LBR tuple, for example, vertex $((A, B), 1, (1, 1))$ corresponds to LBR tuple $((A, B), 1, (1, 1))$. Note that there is edge between vertices $((A, B), 1, (1, 1))$ and $((B, C), 1, (2, 1))$ because (A, B) is incident to (B, C) over band 1. There is an edge between vertices $((A, B), 1, (1, 1))$ and $((B, C), 2, (1, 1))$ because they share a radio in common at CR router B. Similar analysis applies to the other vertices in the conflict graph as well.

3.2.1.2 3-D Independent Sets and Conflict Cliques

Given a 3-D conflict graph $\mathcal{G} = (\mathcal{V}, \mathcal{E})$ representing the CRN, we describe the impact of vertex $i \in \mathcal{V}$ on vertex $j \in \mathcal{V}$ as follows,

$$w_{ij} = \begin{cases} 1, & \text{if there is an edge between vertex } i \text{ and } j \\ 0, & \text{if there is no edge between vertex } i \text{ and } j, \end{cases} \quad (3.4)$$

where two vertices correspond to two LBR tuples, respectively.

Provided that there is a vertex set $\mathscr{I} \subseteq \mathscr{V}$ and an LBR tuple $i \in \mathscr{I}$ satisfying $\sum_{j \in \mathscr{I}, i \neq j} w_{ij} < 1$, the transmission at LBR tuple i will be successful even if all the other LBR tuples in the set $\mathscr{I}$ are transmitting at the same time. If any $i \in \mathscr{I}$ satisfies the condition above, we can schedule the transmissions over all these LBR tuples in $\mathscr{I}$ to be active simultaneously. Such a vertex/LBR tuple set $\mathscr{I}$ is called a 3-D independent set. If adding any one more LBR tuple into a 3-D independent set $\mathscr{I}$ results in a non-independent one, $\mathscr{I}$ is defined as a maximal 3-D independent set. Besides, if there exists a vertex/LBR tuple set $\mathscr{Z} \subseteq \mathscr{V}$ and any two vertexes i and j in $\mathscr{Z}$ satisfying $w_{ij} \neq 0$ (i.e., LBR tuples i and j cannot be scheduled to transmit successfully at the same time.), $\mathscr{Z}$ is called a 3-D conflict clique. If $\mathscr{Z}$ is no longer a 3-D conflict clique after adding any one more LBR tuple, $\mathscr{Z}$ is defined as a maximal 3-D conflict clique.

3.2.2 CR Link Scheduling and Flow Routing Constraints

3.2.2.1 CR Link Scheduling Constraints

Link scheduling can be conducted in time domain, in frequency domain, or in both of them [9, 12]. In this paper, we only focus on time based link scheduling.

Given the 3-D conflict graph $\mathscr{G} = (\mathscr{V}, \mathscr{E})$ constructed from the CRN, suppose we can list all maximal 3-D independent sets[4] as $\mathscr{I} = \{\mathscr{I}_1, \mathscr{I}_2, \cdots, \mathscr{I}_q, \cdots, \mathscr{I}_Q\}$, where Q is $|\mathscr{I}|$, and $\mathscr{I}_q \subseteq \mathscr{V}$ for $1 \leq q \leq Q$. At any time, at most one maximal 3-D independent set can be active to transmit packets for all LBR tuples in that set. Let $\lambda_q \geq 0$ denote the time share scheduled to the maximal 3-D independent set $\mathscr{I}_q$, and

$$\sum_{1 \leq q \leq Q} \lambda_q \leq 1, \quad \lambda_q \geq 0 \, (1 \leq q \leq Q). \tag{3.5}$$

Let $r_{ij}^m(\mathscr{I}_q)$ be the data rate for CR link (i, j) over band m, where $r_{ij}^m(\mathscr{I}_q) = 0$ if LBR tuple $((i, j), m, (u, v)) \notin \mathscr{I}_q$; otherwise, $r_{ij}^m(\mathscr{I}_q)$ is the achievable data rate for CR link (i, j) over band m, which can be calculated from (3.2). Therefore, by exploiting the 3-D maximal independent set $\mathscr{I}_q$, the flow rate that link (i, j) can support over band m in λ_q is $\lambda_q r_{ij}^m(\mathscr{I}_q) T_{ij}^m$.

Furthermore, let $f_{ij}(l)$ represent the flow rate of the session l over link (i, j), where $i \in \mathscr{N}, l \in \mathscr{L}$, and $j \in \bigcup_{m \in \mathscr{M}_i} \mathscr{T}_i^m$. Then, the trading CR sessions are feasible at link (i, j) if there exists a schedule of the maximal 3-D independent sets satisfying

$$\sum_{\substack{l \in \mathscr{L}}}^{s_r(l) \neq j, d_t(l) \neq i} f_{ij}(l)\delta(l) \leq \sum_{q=1}^{|\mathscr{I}|} \lambda_q \sum_{m \in \mathscr{M}_i \cap \mathscr{M}_j} r_{ij}^m(\mathscr{I}_q) T_{ij}^m. \tag{3.6}$$

---

[4]It is a NP-complete problem to find all maximal independent sets in $\mathscr{G}$ [4, 5, 10], which will be further addressed later in this paper. In this section, we make the assumption we could find all the maximal independent sets just for the convenience of our theoretical analysis.

Note that in the equation above, T_{ij}^m is a random variable, which represents the uncertain spectrum supply in the temporal domain as introduced in Sect. 3.1.2. In order to calculate the link capacity achieved by CR link scheduling in (3.6), we need to quantify the temporal spectrum availability when the vacancy of the licensed band is uncertain and modeled as a random variable. Inspired by the mathematical expression of value at risk (VaR) in [8], we leverage parameter α to define temporal spectrum availability at α and denote it by $X_\alpha(T)$ as follows:

$$
\begin{cases}
H_T(\tau) = \displaystyle\int_\tau^\infty h_T(t)dt, & \tau \in \mathscr{R} \\
X_\alpha(T) = \sup\{\tau : H_T(\tau) \geq \alpha\}, & \alpha \in [0, 1].
\end{cases}
\tag{3.7}
$$

Similar to the definition of the X loss in [11] and that of bandwidth integration in [12], $X_\alpha(T)$ can take the best usage of the statistics of T and quantify the temporal spectrum vacancy at confidence level of α. So, based on the definition of temporal spectrum availability at α, Eq. (3.6) can be reformulated as

$$
\sum_{l \in \mathscr{L}}^{s_r(l) \neq j, d_t(l) \neq i} f_{ij}(l)\delta(l) \leq \sum_{q=1}^{|\mathscr{I}|} \lambda_q X_\alpha \Big(\sum_{m \in \mathscr{M}_i \cap \mathscr{M}_j} r_{ij}^m(\mathscr{I}_q) T_{ij}^m \Big).
\tag{3.8}
$$

3.2.2.2 CR Routing Constraints

As for routing, the SSP will help the source CR mesh router to find the available paths and employ a number of relay CR mesh routers to forward the data packets toward its destination CR mesh router. It is obvious that there should be more than one path involved in data delivery since multi-path routing[5] is more flexible to route the traffic from a source router to its destination. Similar to the modeling in [9, 12], we mathematically present routing constraints as follows.

To simplify the notation, let $\mathscr{T}_i = \bigcup_{m \in \mathscr{M}_i} \mathscr{T}_i^m$. If CR mesh router i is the source router of session l, i.e., $i = s_r(l)$, then

$$
\sum_{j \in \mathscr{T}_i} f_{ji}(l) = 0,
\tag{3.9}
$$

$$
\sum_{j \in \mathscr{T}_i} f_{ij}(l)\delta(l) = r(l)\delta(l),
\tag{3.10}
$$

where $\delta(l) \in \{0, 1\}$ indicates whether session l is accepted by the SSP (i.e., session l wins the opportunity for data transmission via spectrum trading) or not.

---

[5]The multiple radios of CR routers allow for multi-path routing.

If CR mesh router i is an intermediate relay router of session l, i.e., $i \neq s_r(l)$ and $i \neq d_t(l)$, then

$$\sum_{j \in \mathcal{T}_i}^{j \neq s_r(l)} f_{ij}(l)\delta(l) = \sum_{p \in \mathcal{T}_i}^{p \neq d_t(l)} f_{pi}(l)\delta(l).$$ (3.11)

If CR mesh router i is the destination router of session l, i.e., $i = d_t(l)$, then

$$\sum_{j \in \mathcal{T}_i} f_{ji}(l)\delta(l) = r(l)\delta(l).$$ (3.12)

Note that if (3.9), (3.10) and (3.11) are satisfied, it can be easily verified that (3.12) must be satisfied. As a result, it is sufficient to list only (3.9), (3.10) and (3.11) as CR routing constraints in CRNs.

3.2.3 Optimal Spectrum Trading Under Multiple Constraints

In order to optimally trade spectrum resources and determine the access/denial of certain CR sessions, the SSP must consider the rate requirements and bidding values of CR sessions, the competition among different CR sessions, the availability of bands (including the SSP's leftover spectrum and the harvested spectrum), and the efficient utilization of spectrum resources. Thus, the SSP seeks for a feasible solution to trading the available frequency bands, assigning these bands to CR mesh routers, scheduling bands for CR transmission and reception, and routing those CR flows so that the revenue of the SSP is maximized and radio spectrum resources are efficiently utilized in multi-hop CRNs.

With the proposed trading system, the optimal spectrum trading problem under multiple constraints in multi-hop CRNs can be formulated as follows:

$$\text{Maximize} \sum_{l \in \mathcal{L}} b(l)\delta(l)$$

$$\text{s.t.:} \quad \sum_{j \in \mathcal{T}_i} f_{ji}(l) = 0 \qquad (l \in \mathcal{L}, i = s_r(l))$$ (3.13)

$$\sum_{j \in \mathcal{T}_i} f_{ij}(l)\delta(l) = r(l)\delta(l) \qquad (l \in \mathcal{L}, i = s_r(l))$$ (3.14)

$$\sum_{j \in \mathcal{T}_i}^{j \neq s_r(l)} f_{ij}(l)\delta(l) = \sum_{p \in \mathcal{T}_i}^{p \neq d_t(l)} f_{pi}(l)\delta(l)$$

$$(l \in \mathcal{L}, i \in \mathcal{N}, i \neq s_r(l), d_t(l))$$ (3.15)

$$\overset{s_r(l)\neq j, d_t(l)\neq i}{\sum_{l\in\mathscr{L}} f_{ij}(l)\delta(l)} \leq \sum_{q=1}^{|\mathscr{I}|}\lambda_q X_\alpha\Big(\sum_{m\in\mathscr{M}_i\cap\mathscr{M}_j} r_{ij}^m(\mathscr{I}_q)T_{ij}^m\Big)$$

$$(i\in\mathscr{N}, j\in\mathscr{T}_i, m\in\mathscr{M}_i\cap\mathscr{M}_j \text{ and } \mathscr{I}_q\in\mathscr{I}) \tag{3.16}$$

$$\sum_{q=1}^{|\mathscr{I}|}\lambda_q \leq 1, \ \lambda_q \geq 0 \qquad (\mathscr{I}_q\in\mathscr{I}) \tag{3.17}$$

$$f_{ij}(l) \geq 0 \ (l\in\mathscr{L}, i\in\mathscr{N}, i\neq d_t(l), j\in\mathscr{T}_i, j\neq s_r(l)) \tag{3.18}$$

$$\delta(l) \in \{0, 1\} \qquad (l\in\mathscr{L}), \tag{3.19}$$

where $\delta(l)$, $f_{ij}(l)$, and λ_q are optimization variables, and $r(l)$ is deterministic value when session l is given. Here, (3.13), (3.14), and (3.15) specify the routing constraints in CRNs. Equations (3.16) and (3.17) indicate that the flow rates over link (i, j) cannot exceed the capacity of this CR link, which is obtained from the CR link scheduling as illustrated in Sect. 3.2.2. Note that $\mathscr{I}$ includes all independent sets in CRNs. Given all the maximal 3-D independent sets[6] in $\mathscr{G}(\mathscr{V}, \mathscr{E})$, we find that the formulated optimization is a mixed-integer linear programming (MILP) problem, which is NP-hard to solve.

3.3 The Upper Bound for the Session Based Spectrum Trading Optimization

The complexity of the optimization above arises from two parts: (1) identifying all the maximal independent sets and (2) fixing the binary $\delta(l)$-variables. To find all the maximal independent sets/cliques itself is NP-complete, but it is not a unique problem in spectrum trading. It has been well investigated in prior multi-hop wireless networks and many approximation algorithms have been proposed in existing literature [10, 16]. For example, one of the typical approaches is to employ K ($0 \leq K \leq |\mathscr{I}|$) maximal independent sets (or a number of maximal conflict cliques) for approximation instead of finding out all the maximal independent sets in $\mathscr{G}(\mathscr{V}, \mathscr{E})$.

On the other hand, $\delta(l)$-variables will be involved as long as the SSP conducts the session based spectrum trading in multi-hop CRNs. Given all the maximal independent sets, we relax the binary requirement on $\delta(l)$ and replace it with $0 \leq \delta(l) \leq 1$ to reduce the complexity for the cross-layer optimization. Due to the enlarged optimization space (caused by relaxation on $\delta(l)$), the solution to

---

[6]That is a general assumption used in existing literature [10, 15, 16] for obtaining throughput bounds or performance comparison.

this relaxed optimization problem yields an upper bound for the SSP's revenue maximization problem. Although the upper bound may not be achieved by a feasible solution, it can play as a benchmark to evaluate the quality of feasible solutions.

3.4 A Bidding Value-Rate Requirement Ratio Based Heuristic Algorithm for Spectrum Trading

In order to find feasible solutions, in this section, we propose a *bidding value-rate requirement ratio* (BVR³) based heuristic algorithm for the SSP's revenue maximization problem. According to the bidding values and rate requirements of candidate trading sessions, we make the SSP classify those CR sessions into different categories in terms of decreasing access possibility. Then, we sequentially fix the $\delta(l)$-variables in different sets and give a heuristic solution, which is also a lower bound for the original MILP problem.

3.4.1 The BVR³ Based Relax-and-Fix Algorithm

The key to simplifying the NP-hard optimization, fixing flow routing (i.e., $f_{ij}(l)$-variables) and link scheduling (i.e., λ_q-variables), and attaining a feasible solution is the determination of the binary values for the $\delta(l)$-variables [9, 12]. Although we can employ the classical branch-and-bound approach to determine $\delta(l)$-variables, the number of iterations involved in that algorithm grows exponentially with $|\mathcal{L}|$. To reduce the complexity, we propose a BVR³ based *relax-and-fix* algorithm [13]. The intuition behind the proposed algorithm is that given the leftover basic spectrum and the harvested spectrum, the SSP would like to take the best use of spectrum resources to make as much revenue as possible. That can be roughly interpreted as the SSP prefers to access the CR session with large bidding value and small rate requirements in spectrum trading. The detailed procedure of the heuristic algorithm for the SSP's revenue maximization is presented as follows.

Based on bidding values and rate requirements of candidate CR sessions, we first sort all the CR sessions in terms of $\frac{b(l)}{r(l)}$ and partition these sessions into S disjoint session sets $\mathcal{L}^1, \mathcal{L}^2, \cdots, \mathcal{L}^S$ in the order of decreasing BVR³, where $\bigcup_{s \in \mathcal{S}} \mathcal{L}^s = \mathcal{L}$ and $\mathcal{S} = \{1, 2, \cdots, S\}$. The BVR³ of the session in $\mathcal{L}^i$ is larger than that of the session in $\mathcal{L}^j$, if i is less than j ($\forall i, j \in \mathcal{S}$).

Then, we create auxiliary session sets by choosing subsets $\mathscr{A}^s$ with $\mathscr{A}^s \subseteq \bigcup_{u=s+1}^{S} \mathcal{L}^u$ for $s \in \{1, 2, \cdots, S-1\}$. For example, in the spectrum trading problem, $\mathcal{L}^1$ may include the $\delta(l)$-variables associated with candidate trading sessions in $\{1, 2, \cdots, l_1\}$, $\mathcal{L}^2$ may be associated with sessions in $\{l_1 + 1, l_1 + 2, \cdots, l_2\}$, and so on, whereas $\mathscr{A}^1$ would include the $\delta(l)$-variables associated with sessions in $\{l_1 + 1, l_1 + 2, \cdots, a_1\}$, and so on.

By leveraging partitioned session sets (i.e., $\mathscr{L}^s$) and auxiliary session sets (i.e., $\mathscr{A}^s$), we sequentially solve $|\mathscr{S}|$ relaxed-MILPs (R-MILPs) (denoted by $R\text{-}MILP^s$ with $1 \leq s \leq |\mathscr{S}|$), determine the δ-variables in $\mathscr{L}^s$ ($s \in \mathscr{S}$) and find a heuristic solution to the original MILP problem. Specifically, in the first R-MILP, $R\text{-}MILP^1$, we only impose the binary requirement on the $\delta(l)$-variables for session l in $\mathscr{L}^1 \cup \mathscr{A}^1$ and relax the integrality restriction on all the other $\delta(l)$-variables for session l in $\mathscr{L}$. Thus, we have

$$R\text{-}MILP^1 \qquad \text{Maximize} \sum_{l \in \mathscr{L}} b(l)\delta(l)$$

$$\text{s.t.:} \quad (3.13), (3.14), (3.15), (3.16), (3.17), (3.18)$$

$$\delta(l) \in \{0, 1\} \qquad (\forall l \in \mathscr{L}^1 \cup \mathscr{A}^1)$$

$$\delta(l) \in [0, 1] \qquad (\forall l \in \mathscr{L}\backslash(\mathscr{L}^1 \cup \mathscr{A}^1)).$$

Let $\{\hat{\delta}^1(1), \cdots, \hat{\delta}^1(l), \cdots, \hat{\delta}^1(L)\}$ be an optimal solution to $R\text{-}MILP^1$. We can fix the $\delta(l)$-variables in $\mathscr{L}^1$ at their corresponding binary values, i.e., $\delta(l) = \hat{\delta}^1(l) \in \{0, 1\}$ for all $l \in \mathscr{L}^1$. Then, we move to $R\text{-}MILP^2$.

In the subsequent $R\text{-}MILP^s$ (for $2 \leq s \leq S$), we sequentially fix the binary values of the $\delta(l)$-variables for sessions in $\mathscr{L}^{s-1}$ from the solution to $R\text{-}MILP^{s-1}$. After that, we further add the binary restriction for the $\delta(l)$-variables in $\mathscr{L}^s \cup \mathscr{A}^s$, and we have

$$R\text{-}MILP^s \qquad \text{Maximize} \sum_{l \in \mathscr{L}} b(l)\delta(l)$$

$$\text{s.t.:} \quad (3.13), (3.14), (3.15), (3.16), (3.17), (3.18)$$

$$\delta(l) = \hat{\delta}^{s-1}(l) \qquad (\forall l \in \mathscr{L}^1 \cup \cdots \cup \mathscr{L}^{s-1})$$

$$\delta(l) \in \{0, 1\} \qquad (\forall l \in \mathscr{L}^s \cup \mathscr{A}^s)$$

$$\delta(l) \in [0, 1] \qquad (\forall l \in \mathscr{L}\backslash(\mathscr{L}^1 \cup \cdots \cup \mathscr{L}^s \cup \mathscr{A}^s)).$$

Either $R\text{-}MILP^s$ is infeasible for certain $s \in \mathscr{S}$ and the heuristic algorithm has failed, or else the proposed BVR[3] based relax-and-fix algorithm provides a feasible solution (i.e., the solution to $R\text{-}MILP^{|\mathscr{S}|}$) to the original MILP problem. The procedure of the proposed heuristic algorithm is summarized in Algorithm 1.

For illustrative purposes, we take a multi-hop CRN consisting of 7 candidate trading CR sessions as an example. We sort these sessions by BVR[3] and divide them into 4 disjoint session sets, i.e., $|\mathscr{S}| = 4$. We conduct the BVR[3] based relax-and-fix algorithm with the following sets $\mathscr{L}^s$ and $\mathscr{A}^s$: $\mathscr{L}^1 = \{1, 2\}$, $\mathscr{L}^2 = \mathscr{A}^1 = \{3, 4\}$, $\mathscr{L}^3 = \mathscr{A}^2 = \{5, 6\}$, and $\mathscr{L}^4 = \mathscr{A}^3 = \{7\}$. The iterations of the heuristic algorithm are as follows:

Algorithm 1 The BVR^3 Based Relax-and-Fix Algorithm

1: Sort all the CR sessions in terms of BVR^3, i.e., $\frac{b(l)}{r(l)}$.
2: Partition all these sessions into S disjoint session sets, denoted by $\mathscr{L}^s$ ($s \in \mathscr{S} = \{1, 2, \cdots, S\}$ and $\mathscr{L}^s \subset \mathscr{L}$).
3: Create auxiliary session sets $\mathscr{A}^s \subseteq \bigcup_{u=s+1}^{S} \mathscr{L}^u$.
4: Set $s = 1$ and relax binary requirement on $\delta(l)$-variables.
5: **for all** $s \in \mathscr{S}$ **do**
6: 　　　Impose binary requirement on the $\delta(l)$-variables for session $l \in \mathscr{L}^s \cup \mathscr{A}^s$.
7: 　　　Using $\mathscr{L}^s$ and $\mathscr{A}^s$, solve the relaxed R-$MILP^s$.
8: 　　　**if** R-$MILP^s$ has a feasible solution **then**
9: 　　　　　Determine the δ-variables in $\mathscr{L}^s$.
10: 　　　　　$s = s + 1$. **continue**
11: 　　**else**
12: 　　　　　Return there is no feasible solution.
13: 　　**end if**
14: **end for**
15: Output the solution to R-$MILP^{|\mathscr{S}|}$ as a feasible solution to the original MILP.

- In the first R-$MILP^1$, the $\delta(l)$-variables associated with sessions in $\{1, \cdots, 4\}$ (i.e., in $\mathscr{L}^1 \cup \mathscr{A}^1$) are restricted to be binary values, the other $\delta(l)$-variables being relaxed.
- From the solution to R-$MILP^1$, we can fix the $\delta(l)$-variables corresponding to the sessions in $\{1, 2\}$ (i.e., in $\mathscr{L}^1$). With the determined $\delta(l)$-variables for sessions in $\mathscr{L}^1$, we continue to solve R-$MILP^2$ where the $\delta(l)$-variables associated with sessions in $\{3, \cdots, 6\}$ (i.e., in $\mathscr{L}^2 \cup \mathscr{A}^2$) are now integer and $\delta(l)$-variables in $\{7\}$ (i.e., in $\mathscr{L} \backslash (\mathscr{L}^1 \cup \mathscr{L}^2 \cup \mathscr{A}^2)$) are relaxed.
- From the solution to R-$MILP^2$, we can additionally fix the $\delta(l)$-variables corresponding to the sessions in $\{3, 4\}$ (i.e., in $\mathscr{L}^2$). Similarly, we can solve R-$MILP^3$ where the $\delta(l)$-variables associated with sessions in $\{5, 6, 7\}$ (i.e., in $\mathscr{L}^3 \cup \mathscr{A}^3$) are now binary and there are no $\delta(l)$-variables to relax because $\mathscr{L} \backslash (\mathscr{L}^1 \cup \mathscr{L}^2 \cup \mathscr{L}^3 \cup \mathscr{A}^3) = \phi$.
- Based on the optimal solution to R-$MILP^3$, we can easily determine the value of $\delta(l)$ in $\{7\}$ and determine whether there is feasible solution to the original MILP.

　　The basic idea of the BVR^3 based relax-and-fix algorithm is explicitly explained in the example. At each iteration, we solve a R-$MILP^s$ problem involving $\mathscr{L}^s \cup \mathscr{A}^s$ sessions and to avoid being too myopic we then only fix the $\delta(l)$-variables corresponding to sessions in $\mathscr{L}^s$. The auxiliary session sets $\mathscr{A}^s$ smooth the heuristic solution by creating some overlap between successive session sets.

　　Different from the upper bound obtained in Sect. 3.3, the proposed BVR^3 based relax-and-fix algorithm yields a lower bound to the optimal spectrum trading problem formulated in Sect. 3.2.3, provided that there exist feasible solutions.

3.4.2 A Coarse-Grained Relax-and-Fix Heuristic Algorithm

Following the same procedure in Sect. 3.3, we first relax the original MILP into LP and find the optimal solution to the relaxed LP, in which $\delta(l)$'s value is in $[0, 1]$. By employing a threshold $0.5 \leq \theta < 1$, we coarsely set the $\delta(l)$-variables exceeding θ to 1 and the other $\delta(l)$-variables to 0. Denote the value of $\delta(l)$ in this solution as $\tilde{\delta}(l) \in \{0, 1\}$. In addition, we keep the same decomposition of session sets as the BVR[3] based relax-and-fix algorithm, i.e., $\mathscr{L}^s$ and $\mathscr{A}^s$ for $s \in \mathscr{S}$.

Then, at each step s ($s \in \mathscr{S}$), all $\delta(l)$-variables are fixed at their $\tilde{\delta}(l)$ values in the best solution found so far (or in the last solution encountered), except the $\delta(l)$-variables in the set $\mathscr{L}^s \cup \mathscr{A}^s$ which are restricted to binary values. Therefore, the problem solved at step s is

$$R\text{-}MILP^s \qquad \text{Maximize} \sum_{l \in \mathscr{L}} b(l)\delta(l)$$

$$\text{s.t.:} \quad (3.13), (3.14), (3.15), (3.16), (3.17), (3.18)$$

$$\delta(l) = \tilde{\delta}(l) \qquad (\forall l \in \mathscr{L} \backslash (\mathscr{L}^s \cup \mathscr{A}^s))$$

$$\delta(l) \in \{0, 1\} \qquad (\forall l \in \mathscr{L}^s \cup \mathscr{A}^s).$$

If a better solution is found, $\tilde{\delta}(l)$ is updated and the fixing procedure continues. Compared with the BVR[3] based relax-and-fix algorithm, different steps s ($s \in \mathscr{S}$) in coarse-grained relax-and-fix heuristic are independent of one another, and any subset of $\mathscr{S}$ can be performed in any order.

3.5 Performance Evaluation

3.5.1 Simulation Setup

We consider a spectrum market in multi-hop CRNs consisting of a SSP, $|\mathscr{N}| = 36$ CR mesh routers and $|\mathscr{L}| = 18$ candidate trading sessions, each of which has a random rate requirement within $[10, 30]$ Mb/s. The bidding values of these sessions are within $[100, 300]$. All CR mesh routers use the same power $P = 10$ W for transmission. Considering the AWGN channel, we assume the noise power η is 10^{-10} W at all routers. Moreover, suppose the path loss factor $\beta = 4$, the antenna parameter $\gamma = 3.90625$, the receiver sensitivity $P_{Tx} = 100\eta = 10^{-8}$ W, and the interference threshold $P_{In} = 6.25 \times 10^{-10}$ W. According to the illustration in Sect. 3.1.2, we can calculate the transmission range R_{Tx} and the interference range R_{In}, which are equal to 250 and 500 m, respectively. For illustrative purposes, we assume all the bands have identical bandwidth, which is set to be 10 MHz, i.e., $W^m = 10$ MHz for all $m \in \mathscr{M}$. Based on the observed data and the statistical analysis in [3], the available time of a licensed band follows the truncated exponential distribution within $[0, 1]$, i.e., $h_T(t, \xi) = \dfrac{\frac{1}{\xi}e^{-\frac{t}{\xi}}}{1-e^{-\frac{1}{\xi}}}$, where $\xi \in (0, 3]$. As for the confidence level, we set $\alpha = 0.85$. Besides, for the simplicity of computation, we set $K = 1 \times 10^4$, i.e., if the total number of the maximal independent sets in $\mathscr{G}(\mathscr{V}, \mathscr{E})$ is less than or equal to 1×10^4, we employ all the maximal independent sets for the solution; otherwise, we employ 1×10^4 maximal independent sets for approximation.

Based on the simulation settings above, we conduct simulations to study the optimal spectrum trading problem in multi-hop CRNs with the following two topologies: (1) a grid topology, where 36 CR mesh routers are distributed within $1000 \times 1000\,\text{m}^2$ area and the area is divided into 25 square cells in $200 \times 200\,\text{m}^2$; (2) a random topology, where 36 CR mesh routers are randomly deployed in a $1000 \times 1000\,\text{m}^2$ area forming a connected network. Note that we employ CPLEX [1] to solve the relaxed optimization problems to obtain the upper bound and lower bounds of the SSP's revenue.

3.5.2 Results and Analysis

In Figs. 3.2 and 3.3, we compare the upper bound of the SSP's revenue with the lower bounds determined by the heuristic BVR^3 based relax-and-fix algorithm (denoted by BRF in figures) and the coarse-grained relax-and-fix algorithm (denoted by CRF in figures) at different number of available bands (i.e., $|\mathscr{M}|$) and radios (i.e., $|\mathscr{H}|$) in multi-hop CRNs. We relax the $\delta(l)$-variables and employ $K = 1 \times 10^4$ maximal independent sets to solve the problem as illustrated in Sect. 3.3, which also yields the upper bound. To develop the lower bounds, we equally divide the 18 candidate trading sessions into 6 session sets (i.e., $|\mathscr{S}| = 6$ and each set has 3 sessions) for the BVR^3 based relax-and-fix algorithm, and set $\theta = 0.7$ for the coarse-grained relax-and-fix algorithm as shown in Sect. 3.4. Given the number of available bands $|\mathscr{M}|$ in CRNs and radios $|\mathscr{H}|$ at CR routers, we employ 50 data sets that can produce feasible solutions and take the average value as a result. For each data set, we re-generate available bands $\mathscr{M}_i$ at CR router i, $s_r(l)/d_t(l)$ and $(r(l), b(l))$ pair of session l, and the random network topology (we keep the same grid topology for each data set), which follows the guideline of simulation setup.

From the results shown in Figs. 3.2 and 3.3, four observations can be made in order. First, the upper bound is close to the lower bounds obtained from the proposed BVR^3 based relax-and-fix algorithm and the coarse-grained relax-and-fix algorithm, no matter how many available bands and radios are there in CRNs. We will further present the ratio of the upper bound to lower bounds with 50 data sets in Fig. 3.4, analyze the statistical results and show the closeness between those bounds. Second, as the number of available bands and the number of CR mesh router's radios increase, the SSP's revenue increases as well. The reason is that more bands and

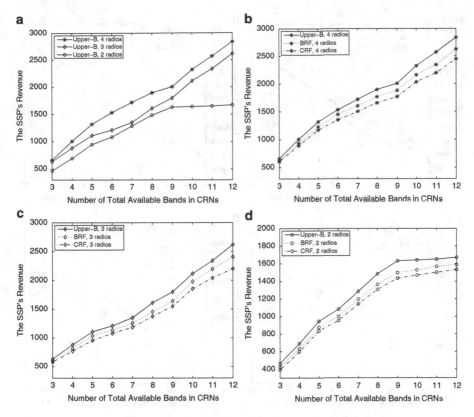

Fig. 3.2 Impact of the number of available bands $|\mathcal{M}|$ and radio interfaces $|\mathcal{H}|$ on spectrum trading in multi-hop CRNs: grid topology. (**a**) Revenue upper bounds: $|\mathcal{H}| = 2, 3$, and 4. (**b**) Revenue upper bound and lower bounds: $|\mathcal{H}| = 4$. (**c**) Revenue upper bound and lower bounds: $|\mathcal{H}| = 3$. (**d**) Revenue upper bound and lower bounds: $|\mathcal{H}| = 2$

radios available create more LBR tuples, so that more CR links may be activated for transmission simultaneously and more opportunities can be leveraged for spectrum trading in CRNs. However, the increment of the SSP's revenue basically stops when $|\mathcal{M}|$ is over 9 for $|\mathcal{H}| = 2$ case in both grid topology and random topology, which leads to the third observation. That is, the CR mesh router has to equip a reasonable number of radios to utilize all the available bands efficiently (at least 3 radios for our simulation scenarios). This observation also gives a good suggestion on the design and deployment of CR mesh routers for spectrum trading in practice. Fourth, the performance of the grid topology generally outperforms that of the random topology in terms of the SSP's revenue. The performance gap stems from the differences in topological structure. For the grid topology, each CR link has the same topological information if we ignore the border effect. The performance improvement of spectrum trading is mainly determined by the number of radios and the available bands at different CR routers. By contrast, the random topology

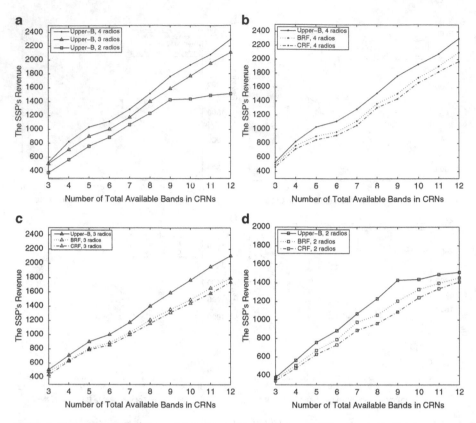

Fig. 3.3 Impact of the number of available bands $|\mathcal{M}|$ and radio interfaces $|\mathcal{H}|$ on spectrum trading in multi-hop CRNs: random topology. (**a**) Revenue upper bounds: $|\mathcal{H}| = 2, 3$, and 4. (**b**) Revenue upper bound and lower bounds: $|\mathcal{H}| = 4$. (**c**) Revenue upper bound and lower bounds: $|\mathcal{H}| = 3$. (**d**) Revenue upper bound and lower bounds: $|\mathcal{H}| = 2$

is non-uniformed topology. The performance improvement of spectrum trading is not only hindered by the number of bands and radios but also bottlenecked by the critical cliques in the random topology.

Figure 3.4 presents the ratio of the upper bound to the lower bounds obtained from the proposed heuristic algorithms in both grid topology and random topology, where $|\mathcal{H}| = 3$ and $|\mathcal{M}| = 9$. As shown in Fig. 3.4a, b, the ratio of the upper bound to lower bound in the grid topology is near to 1 with 50 different data sets, where the lower bounds are determined by the BVR^3 based relax-and-fix algorithm and the coarse-grained relax-and-fix algorithm, respectively. Specifically, the mean ratio of the upper bound to the BVR^3 based lower bound for all the data sets is 1.0973, and the standard deviation is 0.0707; the mean ratio of the upper bound to the coarse-grained based lower bound for all the data sets is 1.1462, and the standard deviation is 0.1255. Similar analysis applies to the random topology as well. As shown in Fig. 3.4c, d, the mean ratio of the upper bound to the BVR^3 based lower bound for

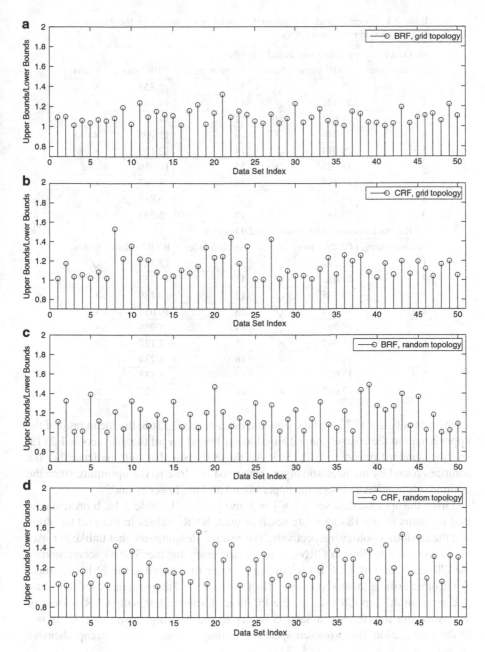

Fig. 3.4 Ratio of the upper bound to lower bounds determined by the proposed algorithms at $|\mathcal{H}| = 3$ and $|\mathcal{M}| = 9$. (**a**) Ratio of the upper bound to the lower bound determined by the BVR³ based relax-and-fix algorithm: grid topology. (**b**) Ratio of the upper bound to the lower bound determined by the coarse-grained relax-and-fix algorithm: grid topology. (**c**) Ratio of the upper bound to the lower bound determined by the BVR³ based relax-and-fix algorithm: random topology. (**d**) Ratio of the upper bound to the lower bound determined by the coarse-grained relax-and-fix algorithm: random topology

Table 3.1 Spectrum trading status of the candidate sessions w.r.t. the descending BVR^3 values in multi-hop CRNs

(a) Grid topology with 3 radios and 9 bands

Session index	BVR^3 value	Status	Session index	BVR^3 value	Status
1	25.001	√	10	15.353	×
2	23.422	√	11	14.742	×
3	21.811	√	12	14.071	×
4	21.489	√	13	12.996	×
5	20.014	√	14	12.159	×
6	19.125	√	15	10.016	√
7	17.475	×	16	8.287	×
8	17.212	×	17	6.883	×
9	16.135	×	18	5.295	×

(b) Random topology with 3 radios and 9 bands

Session index	BVR^3 value	Status	Session index	BVR^3 value	Status
1	24.211	√	10	12.587	√
2	22.023	√	11	11.233	×
3	20.835	√	12	10.489	×
4	19.333	√	13	10.038	×
5	18.025	√	14	8.955	×
6	15.667	×	15	7.122	×
7	14.511	×	16	6.734	×
8	13.936	×	17	5.533	×
9	13.012	×	18	4.327	×

all the data sets is 1.1722, and the standard deviation is 0.1365; the mean ratio of the upper bound to the coarse-grained based lower bound for all the data sets is 1.2113, and the standard deviation is 0.1595. All these statistical results indicate that the solutions found by the heuristic algorithms must be close to the optimum, since the optimal solution lies between the upper bound and the lower bound.

Given the specific data set at $|\mathcal{H}| = 3$ and $|\mathcal{M}| = 9$, Table 3.1a, b presents the trading status of the 18 candidate sessions w.r.t. BVR^3 values in the grid topology and the random topology, respectively. The results[7] demonstrate that unlike per-user based spectrum trading in CRNs, it is not necessary for the SSP to accommodate the CR sessions with high BVR^3 values in order to maximize the SSP's revenue. Some other critical factors may also affect the results of the session based spectrum trading in multi-hop CRNs, e.g., the location of source/destination CR routers of a session, the interference a session incurs to the existing flows, etc. As shown in the formulation, the proposed spectrum trading scheme gives a comprehensive

---

[7]We exploit the proposed BVR^3 based relax-and-fix algorithm to derive these results in both the grid topology and the random topology.

consideration on those factors. The data in Table 3.1 further verify this statement and explicitly show the advantages of our design over the per-user based spectrum trading systems in multi-hop CRNs.

References

1. IBM ILOG CPLEX Optimizer.
2. IEEE 802.22-2011(TM) Standard for Cognitive Wireless Regional Area Networks (RAN) for Operation in TV Bands, July 2011.
3. D. Chen, S. Yin, Q. Zhang, M. Liu, and S. Li. Mining spectrum usage data: a large-scale spectrum measurement study. In *Proc. of international conference on Mobile computing and networking, ACM Mobicom, 2009*, Beijing, China, September 2009.
4. R. Diestel. *Graph Theory*. Springer, 2005.
5. M. R. Garey and D. S. Johnson. *Computers and Intractability: A Guide to the Theory of NP-Completeness*. W. H. Freeman and Company, New York, NY, 1979.
6. A. Goldsmith. *Wireless Communications*. Cambridge University Press, Cambridge, NY, 2005.
7. P. Gupta and P. R. Kumar. The capacity of wireless networks. *IEEE Transactions on Information Theory*, 46(2):388–404, March 2000.
8. G. Holton. *Value-at-Risk: Theory and Practice*. Academic Press, 2003.
9. Y. T. Hou, Y. Shi, and H. D. Sherali. Spectrum sharing for multi-hop networking with cognitive radios. *IEEE Journal on Selected Areas in Communications*, 26(1):146–155, January 2008.
10. H. Li, Y. Cheng, C. Zhou, and P. Wan. Multi-dimensional conflict graph based computing for optimal capacity in MR-MC wireless networks. In *Proc. of International Conference on Distributed Computing Systems, ICDCS 2010*, Genoa, Italy, June 2010.
11. M. Pan, H. Yue, Y. Fang, and H. Li. The x loss: Band-mix selection for opportunistic spectrum accessing with uncertain supply from primary service providers. *IEEE Transactions on Mobile Computing*, 11(12):2133–2144, December 2012.
12. M. Pan, C. Zhang, P. Li, and Y. Fang. Joint routing and scheduling for cognitive radio networks under uncertain spectrum supply. In *Proc. of IEEE Conference on Computer Communications, INFOCOM 2011*, Shanghai, China, April 2011.
13. Y. Pochet and L. A. Wolsey. *Production Planning by Mixed Integer Programming*. Springer-Verlag New York, Inc., Secaucus, NJ, USA, 2006.
14. Y. Shi, Y. T. Hou, and S. Kompella. How to correctly use the protocol interference model for multi-hop wireless networks. In *Proc. of ACM International Symposium on Mobile Ad Hoc Networking and Computing, ACM MobiHoc, 2009*, New Orleans, LA, May 2009.
15. J. Tang, S. Misra, and G. Xue. Joint spectrum allocation and scheduling for fair spectrum sharing in cognitive radio wireless networks. *Computer Networks (Elsevier) Journal*, 52(11):2148–2158, August 2008.
16. H. Zhai and Y. Fang. Impact of routing metrics on path capacity in multirate and multihop wireless ad hoc networks. In *Proc. of the IEEE International Conference on Network Protocols, ICNP 2006*, Santa Barbara, CA, November 2006.

Chapter 4
Economic-Robust Session Based Spectrum Trading

Abstract This chapter further extends the session based spectrum trading into an economic-robust session based one. Beyond considering the end-to-end performance of spectrum trading as illustrated in last chapter, the economic-robust session based spectrum trading also guarantee the economic properties of spectrum trading such as incentive compatibility, individual rationality, and budget balance. By employing two bidding manners, i.e., bidding for the whole session and unit rate bidding, we formulate the spectrum trading problems under multiple economic and multi-hop CR transmission constraints, design two pricing mechanisms to charge the winning spectrum bidders, and further mathematically prove the economic-robustness of the proposed spectrum trading schemes. Through extensive simulations, we show the proposed schemes are economic-robust and effective in improving spectrum utilization.

Keywords Cognitive radio sessions • Economic robustness • Critical value • Pricing mechanism

4.1 Network Model

4.1.1 System Architecture for Spectrum Trading

Similar to the network configuration in last chapter, assume that $\mathcal{N} = \{1, 2, \cdots, N\}$ CR mesh routers are deployed by the SSP and L SUs bid for the resources. Each SU $l \in \mathcal{L} = \{1, 2, \cdots, L\}$ has one session with certain source and destination CR router, rate requirement and bidding value denoted as $s(l)$, $d(l)$, $r(l)$, and $b(l)$, respectively. The CR mesh routers can use the basic bands owned by the SSP, and opportunistically access some non-active licensed bands as well. Specifically, to get the information of all sessions for the trading, some basic bands are utilized. Then the available bands for traffic delivery in the CR router network are the rest of basic bands and the sensed unused licensed bands, denoted as $\mathcal{M} = \{1, 2, \cdots, M\}$, with different bandwidths $\mathcal{W} = \{W_1, W_2, \cdots, W_M\}$. Considering the geographical constraints, each CR router may be not able to access all the available bands and we denote the available bands at the ith CR router as $\mathcal{M}_i \subseteq \mathcal{M}$. Different CR router may

M. Pan et al., *Spectrum Trading in Multi-Hop Cognitive Radio Networks*,
SpringerBriefs in Electrical and Computer Engineering,
DOI 10.1007/978-3-319-25631-3_4

have different available band set, i.e., $\mathcal{M}_i \neq \mathcal{M}_j$, $i \neq j$, and the common available band set between the ith and jth CR router is denoted as $\mathcal{M}_{i \cap j} = \mathcal{M}_i \cap \mathcal{M}_j$.

4.1.2 Related Models in Multi-Hop CRNs

Transmission Range and Interference Range For the power propagation gain from CR router i to j, $i \neq j \in \mathcal{N}$, we adopt a widely used model [2–4, 6–8] shown as

$$g_{ij} = \beta \cdot d_{ij}^{-\alpha}, \tag{4.1}$$

where β is an antenna related parameter, α is the path loss factor, and d_{ij} represents the distance between the two CR routers. Assume that the transmitted power at the ith CR router is P_i, and its data transmission is successful only when the received power can exceed a power threshold as P_{th}^T, i.e., $P_i \cdot g_{ij} \geq P_{th}^T$. Thus, we can obtain the transmission range of the ith CR router as $R_i^T = \left(\beta \cdot P_i / P_{th}^T\right)^{1/\alpha}$. Similarly, suppose that the received interference can be ignored only when its power is less than a threshold as P_{th}^I. Therefore, the interference range of the ith CR router can be denoted as $R_i^I = \left(\beta \cdot P_i / P_{th}^I\right)^{1/\alpha}$. Since $P_{th}^T > P_{th}^I$, for the ith CR router, it is obvious that $R_i^T < R_i^I$.

Link Capacity Assume that the CR router j is in the transmission range of the CR router i and they have a common available band set, i.e., $\mathcal{M}_{i \cap j} \neq \emptyset$. Then, based on the Shannon–Hartley theorem, the link capacity from i to j with band $m \subseteq \mathcal{M}_{i \cap j}$ can be expressed as

$$c_{ij}^m = W^m \log_2 \left(1 + \frac{P_i \cdot g_{ij}}{\gamma}\right), \tag{4.2}$$

where γ is the Gaussian noise power at the CR router j. Interferences are not considered here since they can be handled following the scheduling of the SSP according to the interference range of each CR router. The link capacity is an important constraint for the design of flow routing since the aggregate flow rate on one link cannot exceed its capacity.

4.1.3 Preliminaries for Spectrum Trading

Before the design for session based spectrum trading, we introduce a set of notations and some important economic characteristics in this section.

Bidding Value In this chapter, we consider two manners of bidding. One is for the whole session and the per-session bidding value of SU $l \in \mathcal{L}$ is denoted as $b_s(l)$. The other one is for the unit rate and that based on per-rate is expressed as $b_r(l)$.

True Value For the bid, each SU $l \in \mathscr{L}$ has an own valuation, i.e., the true price they will to pay, which are denoted as $v_s(l)$ and $v_r(l)$ for the two manners, respectively.

Clearing Price According to the bidding values, the auctioneer will decide winners and allocate its resources. Meanwhile, it will charge price for each winner t, denoted as $p_s(t)$ and $p_r(t)$, for one session and unit rate, respectively, corresponding to the two manners of bid.

Bidder Utility For any SU $l \in \mathscr{L}$, the utility functions for the two manners are $u_{s/r}\left(l, b_{s/r}(l)\right) = v_{s/r}(l) - p_{s/r}(l)$ if it wins with bid $b_{s/r}(l)$, and 0 otherwise.

To maintain the stability of trading market, the trading scheme should be economic-robust, i.e., satisfy the following three important economic characteristics:

Incentive Compatibility (IC) A trading scheme is IC if no one can get higher utility by bidding untruthfully no matter how other bidders bid. Mathematically, for bidder i, when others' bids are fixed, $u_{s/r}\left(i, v_{s/r}(i)\right) \geq u_{s/r}\left(i, b_{s/r}(i)\right)$ if $b_{s/r}(i) \neq v_{s/r}(i)$.

Individual Rationality (IR) A trading scheme is IR if the clearing price of any bidder i is not higher than its bidding value, i.e., $p_{s/r}(i) \leq b_{s/r}(i)$.

Budget Balanced (BB) A trading scheme is BB if the generated revenue of the auctioneer is non-negative.

Here, we do not consider the cost at the spectrum trader, SSP, during the trading and thus the revenue is the sum of the clearing price charging for the winning sessions, which is always non-negative. Therefore, BB can be always satisfied in our scheme and we will focus on the other two properties.

4.2 Optimal Resource Allocation for Session Based Spectrum Trading in Multi-Hop CRNs

In this section, we formulate the resource allocation of the session based spectrum trading into an optimization problem to maximize the expected revenue of the SSP under interference constraints and flow routing in the CR router network.

4.2.1 Interference Constraints

In the CR router network, the available spectrum bands should be allocated carefully to avoid interference among different links. We exploit a binary value to describe the condition of the link from router i to j, $i \neq j \in \mathscr{N}$, on band $m \in \mathscr{M}_{i \cap j}$ as

$$x_{ij}^m = \begin{cases} 1, & \text{if } i \text{ can transmit data to } j \text{ with band } m, \\ 0, & \text{otherwise.} \end{cases} \tag{4.3}$$

Furthermore, we denote the set of CR routers, which are in the transmission range of CR router $i \in \mathcal{N}$ and can use band $m \in \mathcal{M}_i$, as

$$\mathcal{T}_i^m = \left\{ j \mid d_{ij} \leq R_i^T, j \neq i, m \in \mathcal{M}_{i \cap j} \right\}. \tag{4.4}$$

Similarly, the CR routers which can interfere with CR router i on band m are expressed as

$$\mathcal{I}_i^m = \left\{ k \mid d_{ki} \leq R_k^I, k \neq i, m \in M_{i \cap k}, \mathcal{T}_k^m \neq \emptyset \right\}. \tag{4.5}$$

Based on the aforementioned illustration, we present the interference constraints. For any CR router $i \in \mathcal{N}$, it cannot transmit to or receive from different routers on the same band and we achieve the constraint C1 as

$$C1: \sum_{j \in T_i^m} x_{ij}^m \leq 1 \text{ and } \sum_{\{i \mid j \in T_i^m\}} x_{ij}^m \leq 1. \tag{4.6}$$

Besides, one CR router cannot transmit and receive on the same band simultaneously considering the "self-interference" at physical layer, which brings the constraint C2 as

$$C2: x_{ij}^m + \sum_{q \in \mathcal{T}_j^m} x_{jq}^m \leq 1. \tag{4.7}$$

Moreover, interference among different CR routers should be noticed as well. According to (4.5), we note that when CR router $i \in \mathcal{N}$ is transmitting data on band $m \in \mathcal{M}_i$, any other routers who can interfere with router i cannot use this band. Thus we can obtain the constraint C3 as

$$C3: x_{ij}^m + \sum_{q \in \mathcal{T}_k^m} x_{kq}^m \leq 1, k \in \mathcal{I}_j^m, k \neq i. \tag{4.8}$$

4.2.2 Flow Routing Constraints

After the design for interference management, how to deliver the traffic for winning sessions using a set of paths from their sources to destinations is also an important issue. In this section, we will present the constraints considered for the flow routing design.

Similarly, we employ a binary variable to denote whether the session $l \in \mathcal{L}$ wins or not as

$$w(l) = \begin{cases} 1, & \text{if session } l \text{ wins the bid}, \\ 0, & \text{otherwise}. \end{cases} \tag{4.9}$$

Let $f_{ij}^m(l)$ represent the flow attributed to session $l \in \mathcal{L}$ on link i to j, $i \neq j \in \mathcal{N}$, using band $m \in \mathcal{M}_{i \cap j}$.

First, consider the source router of the winning session l, i.e., $i = s(l)$. The incoming data should be zero and the sum rate of outgoing transmission to different routers using different bands should meet the rate requirement of session l. Therefore, we have the following constraints as

$$F1: \sum_{m \in \mathcal{M}_{i \cap j}} \sum_{j \in \mathcal{T}_i^m} f_{ji}^m(l) w(l) = 0, i = s(l), \tag{4.10}$$

$$\sum_{m \in M_{i \cap j}} \sum_{j \in T_i^m} f_{ij}^m(l) w(l) = r(l) w(l), i = s(l). \tag{4.11}$$

Next, consider the intermediate routers of the winning session l, i.e., $i \neq s(l)$, $i \neq d(l)$. The total outgoing data rate should be equal to its total incoming data rate to keep the flow balance, which leads to the following constraint as

$$F2: \sum_{m \in \mathcal{M}_{p \cap i}} \sum_{i \in \mathcal{T}_p^m}^{p \neq d(l)} f_{pi}^m(l) w(l) = \sum_{m \in \mathcal{M}_{i \cap j}} \sum_{j \in \mathcal{T}_i^m}^{j \neq s(l)} f_{ij}^m(l) w(l),$$

$$i \neq s(l), i \neq d(l). \tag{4.12}$$

For the destination router of the winning session l, i.e., $i = d(l)$, in contrast to F1, there is no outgoing data and the total incoming data rate should be the rate requirement $r(l)$. Then we have

$$F3: \sum_{m \in M_{i \cap j}} \sum_{j \in T_i^m} f_{ij}^m(l) w(l) = 0, i = d(l), \tag{4.13}$$

$$\sum_{m \in M_{i \cap j}} \sum_{j \in T_i^m} f_{ji}^m(l) w(l) = r(l) w(l), i = d(l). \tag{4.14}$$

Furthermore, considering the link from router i to j, $i \neq j \in \mathcal{N}$, if it is active under the interference constraints, i.e., $\exists x_{ij}^m = 1$, $m \in \mathcal{M}_{i \cap j}$, the sum flow of all winning sessions on this link should not be higher than its capacity, i.e.,

$$F4: \sum_{m \in \mathcal{M}_{i \cap j}} \sum_{l \in L}^{i \neq d(l), j \neq s(l)} f_{ij}^m(l) w(l) \leq \sum_{m \in \mathcal{M}_{i \cap j}} c_{ij}^m x_{ij}^m, j \in \mathcal{T}_i^m, \tag{4.15}$$

where c_{ij}^m is the capacity of link i to j on band m and can be calculated as (4.2).

4.2.3 Problem Formulation

In terms of the spectrum trading system, the objective of the SSP is to maximize its expected revenue under the aforementioned multiple constraints. Two bidding manners are considered. One is that each bidder (session) $l \in \mathscr{L}$ bids for its whole session, i.e., $b_s(l)$, and the optimal resource allocation for the session based spectrum trading in multi-hop CRNs can be formulated as follows:

$$\text{P1: Maximize} \quad \sum_{l \in \mathscr{L}} b_s(l) w(l)$$

$$\text{s.t.} \quad (6) \sim (8) , \quad (10) \sim (15)$$

$$x_{ij}^m, w(l) \in \{0, 1\} \left(i \in \mathscr{N}, j \in \mathscr{T}_j^m, m \in \mathscr{M}_{i \cap j}, l \in \mathscr{L} \right) \tag{4.16}$$

$$f_{ij}^m(l) \geq 0$$

$$\left(l \in \mathscr{L}, i \in \mathscr{N}, i \neq d(l), j \in \mathscr{T}_j^m, j \neq s(l), m \in \mathscr{M}_{i \cap j} \right), \tag{4.17}$$

where $b_s(l)$, $r(l)$, and c_{ij}^m are given constants, and $w(l)$, x_{ij}^m, and $f_{ij}^m(l)$ are optimization variables.

The other bidding manner is that each bidder $l \in \mathscr{L}$ bids for unit rate denoted as $b_r(l)$. Considering the rate requirement of session l expressed as $r(l)$, the sum bidding value of session l actually is $r(l) \cdot b_r(l)$. Thus, the optimal resource allocation is turned to be P2 as follows:

$$\text{P2: Maximize} \quad \sum_{l \in \mathscr{L}} r(l) b_r(l) w(l)$$

$$\text{s.t.} \quad (6) \sim (8) , \quad (10) \sim (17)$$

We can find that the formulated problems, P1 and P2, are MINLP problems, which are hardly to solve. Thus, we substitute $f_{ij}^m(l) \cdot w(l)$ by $F_{ij}^m(l)$, and the problems will turn to be MILP problems, which can be solved by LP_SOLVE.

4.3 Economic-Robust Pricing Mechanism for Session Based Spectrum Trading

According to the bidding values of all sessions and multiple constraints of interference management and flow routing, winning sessions can be decided by the SSP through solving P1 or P2. After that, the SSP will decide the clearing price charging for the winners, which should guarantee IC [9, 10] and IR [5]. In this section, we first present the pricing mechanisms corresponding to P1 and P2, respectively. Then we give the proof of their economic-robustness.

4.3.1 Pricing Mechanism

Consider the first bidding manner corresponding to P1. Denote the set of all winning sessions ($w(l) = 1$) after solving P1 as $\mathscr{L}_1^*$. The clearing price of one session for winner $t \in \mathscr{L}_1^*$ expressed as $p_s(t)$ is defined as follows.

Pricing for One Session Let $S_s(\mathscr{L}_1^*)$ and $S_s^t(\mathscr{L}_1^*)$ denote the total bidding values of all winning sessions and that except winner t, respectively, i.e.,

$$S_s(\mathscr{L}_1^*) = \sum_{l \in \mathscr{L}_1^*} b_s(l) \text{ and } S_s^t(\mathscr{L}_1^*) = \sum_{l \in \mathscr{L}_1^*, l \neq t} b_s(l). \tag{4.18}$$

Assume that the session t quits the bid, i.e., $b_s(t) = 0$, and the set of updated winning sessions through solving P1 is expressed as $\widetilde{\mathscr{L}_1^*}$, and the total bidding value of all updated winning sessions can be described as $S_s(\widetilde{\mathscr{L}_1^*})$. Then we give the clearing price of one session charging for winner $t \in \mathscr{L}_1^*$ after the optimal resource allocation based on P1 as

$$p_s(t) = S_s(\widetilde{\mathscr{L}_1^*}) - S_s^t(\mathscr{L}_1^*). \tag{4.19}$$

Next, we consider the other bidding manner corresponding to P2. Similarly, the set of all winning sessions by solving P2 is denoted as $\mathscr{L}_2^*$ and the clearing price of unit rate for winning session $t \in \mathscr{L}_2^*$ denoted as $p_r(t)$ is defined as follows.

Pricing for Unit Rate Similar to (4.18), the total bids of all winning sessions after solving P2 can be expressed as

$$S_r(\mathscr{L}_2^*) = \sum_{l \in \mathscr{L}_2^*} r(l) b_r(l), \tag{4.20}$$

and the total winning bids without session t is

$$S_r^t(\mathscr{L}_2^*) = \sum_{l \in \mathscr{L}_2^*, l \neq t} r(l) b_r(l). \tag{4.21}$$

Then let $b_r(t) = 0$, and the set of updated winning sessions by solving P2 can be described as $\widetilde{\mathscr{L}_2^*}$, and the updated total bids can be denoted as $S_r(\widetilde{\mathscr{L}_2^*})$. Then the clearing price of unit rate charging for winner $t \in \mathscr{L}_2^*$ after the optimal resource allocation based on P2 is set as

$$p_r(t) = \frac{S_r(\widetilde{\mathscr{L}_2^*}) - S_r^t(\mathscr{L}_2^*)}{r(t)}. \tag{4.22}$$

4.3.2 Proof of Economic-Robustness

In this section, we will prove that our trading scheme with two bidding manners is IR and IC.

Firstly, We focus on the bidding manner for one session, in which the resource allocation and pricing mechanism correspond to P1 and (4.19), respectively.

Theorem 4.1. *The proposed trading scheme with bidding for one session is IR.*

Proof. Since $\mathscr{L}_1^*$ is the winner set of P1, the value of the objective function corresponding to $\mathscr{L}_1^*$, i.e., the total bids of all winning sessions, should be maximum. Thus, we have

$$S_s\left(\mathscr{L}_1^*\right) \geq S_s\left(\widetilde{\mathscr{L}_1^*}\right). \tag{4.23}$$

Then, for each winner $t \in \mathscr{L}_1^*$, we can get

$$b_s(t) = S_s\left(L_1^*\right) - S_s^t\left(L_1^*\right) \geq S_s\left(\widetilde{L_1^*}\right) - S_s^t\left(L_1^*\right) = p_s(t), \tag{4.24}$$

which means that the IR property can be satisfied.

Before giving the proof of IC, we first present some definitions and derive some lemmas.

Definition 4.1 (Monotonic Allocation). For any bidder l, when the bids of other bidders are fixed, if it can win the resources with bidding value $b(l)$, then it can always win by bidding $\overline{b(l)} \geq b(l)$. On the contrary, if it loses with $b(l)$, it will always lose by bidding $\underline{b(l)} \leq b(l)$.

Definition 4.2 (Critical Value). Critical value is a boundary value. For any bidder l, if it bids higher than its critical value, it will win, and if it bids lower than that, it will lose.

Lemma 4.1. *The resource allocation with bidding for one session of our trading scheme as P1 is a monotonic allocation.*

Proof. We prove it by contradiction. Considering any winner $t \in \mathscr{L}_1^*$ with bidding value $b_s(t)$, and others' bids are fixed, we make an assumption that if it bids higher with $\overline{b_s(t)} \geq b_s(t)$, it would lose. We denote the new set of winning sessions as $\overline{\mathscr{L}_1^*}$, $\overline{\mathscr{L}_1^*} \neq \mathscr{L}_1^*$ and $t \notin \overline{\mathscr{L}_1^*}$. Then we have

$$\sum_{l \in \overline{L_1^*}} b_s(l) \geq \sum_{l \in \mathscr{L}_1^*, l \neq t} b_s(l) + \overline{b_s(t)} \geq \sum_{l \in \mathscr{L}_1^*, l \neq t} b_s(l) + b_s(t). \tag{4.25}$$

We can find that the original set of winners by solving P1 when bidder t bids $b_s(t)$ should be $\overline{\mathscr{L}_1^*}$ other than $\mathscr{L}_1^*$. In other words, bidder t loses with bidding value

$b_s(t)$ which is the contradiction. Therefore, when the winner t bids $\overline{b_s(t)} \geq b_s(t)$, it must win as well. On the other hand, for any loser $z \notin \mathscr{L}_1^*$ with bid $b_s(z)$, it can be proved in the similar way that it also loses if its bid is $\overline{b_s(z)} \leq b_s(z)$.

Lemma 4.2. *The clearing price $p_s(t)$ for each winner $t \in \mathscr{L}_1^*$ is a critical value.*

Proof. (1) Assume that bidder t loses with $b_s(t) > p_s(t)$. Then the set of winners should be $\widetilde{\mathscr{L}_1^*}$. According to (4.19), we have

$$S_s\left(\mathscr{L}_1^*\right) = S_s^t\left(\mathscr{L}_1^*\right) + b_s(t) > S_s^t\left(\mathscr{L}_1^*\right) + p_s(t) = S_s\left(\widetilde{\mathscr{L}_1^*}\right), \qquad (4.26)$$

which indicates that the winner set should not be $\widetilde{\mathscr{L}_1^*}$ and bidder t should be a winner. Thus we have a contradiction and bidder t can win if it bids $b_s(t) > p_s(t)$. (2) Assume that bidder t wins with $b_s(t) < p_s(t)$. Then the set of winners should be $\mathscr{L}_1^*$. Similar to (4.26), we can get $S_s\left(\mathscr{L}_1^*\right) < S_s\left(\widetilde{\mathscr{L}_1^*}\right)$, which means that bidder t should be a loser and the contradiction is also achieved. Thus bidder t will lose if it bids $b_s(t) < p_s(t)$. Overall the clearing price is a critical value.

Based on Lemmas 4.1 and 4.2, we can prove the satisfaction of IC and have the following theorem.

Theorem 4.2. *The proposed trading scheme with bidding for one session is IC.*

Proof. According to the definition of IC, we will show that for any bidder l, when others' bids are fixed, $u_s(l, v_s(l)) \geq u_s(l, b_s(l))$ if $b_s(l) \neq v_s(l)$.

We start from $b_s(l) > v_s(l)$ and discuss all four possible cases as follows:

> *Case 1:* Bidder l loses with both bids. In this case, $u_s(l, v_s(l)) = u_s(l, b_s(l)) = 0$ and the claim holds.
> *Case 2:* Bidder l wins with $v_s(l)$ but loses with $b_s(l)$. From Lemma 4.1, this case cannot happen.
> *Case 3:* Bidder l loses with $v_s(l)$ but wins with $b_s(l)$. According to Lemma 4.2, we can obtain that $v_s(l) < p_s(l) < b_s(l)$. Thus $u_s(l, b_s(l)) = v_s(l) - p_s(l) < 0$ but $u_s(l, v_s(l)) = 0$. Hence the claim holds.
> *Case 4:* Bidder l wins with both bids. Since the constraints in P1 are independent of bidding values, the sets of winners should be the same when bidder l bids $v_s(l)$ and $b_s(l)$ but others' bids are fixed. Then, according to (4.19), we find that the clearing price for winner t will be equal for the two scenarios and thus $u_s(l, v_s(l)) = u_s(l, b_s(l))$.

When $b_s(l) < v_s(l)$, the proofs for the four cases are similar and thus omitted here due to the limited space.

Overall, we reach the conclusion that $u_s(l, v_s(l)) \geq u_s(l, b_s(l))$ if $b_s(l) \neq v_s(l)$, and thus the IR property of our scheme is proved.

Next, we consider the other bidding manner for unit rate, where the resource allocation and pricing mechanism are P2 and (4.22), respectively, and give the following two theorems.

Theorem 4.3. *The proposed trading scheme with bidding for unit rate is IR.*

Proof. Similar to the proof of Theorem 4.2, considering that $\mathscr{L}_2^*$ is the set of winners of P2, we have

$$S_r\left(\mathscr{L}_2^*\right) \geq S_s\left(\widetilde{\mathscr{L}_2^*}\right). \tag{4.27}$$

Then for any winner $t \in \mathscr{L}_2^*$, we can get

$$b_r(t) = \frac{S_r\left(\mathscr{L}_2^*\right) - S_r^t\left(\mathscr{L}_2^*\right)}{r(t)} \geq \frac{S_r\left(\widetilde{\mathscr{L}_2^*}\right) - S_r^t\left(\mathscr{L}_2^*\right)}{r(t)} = p_r(t), \tag{4.28}$$

and thus the IR property can be guaranteed.

Theorem 4.4. *The proposed trading scheme with bidding for unit session is IC.*

Proof. The resource allocation P2 is also a monotonic allocation as P1 and the clearing price $p_r(t)$ for each winner $t \in \mathscr{L}_2^*$ is also a critical value as $p_s(t)$, which can be proved similar to Lemmas 4.1 and 4.2. Then the theorem can be proved by discussing the cases similar to the proof of Theorem 4.2. The details are omitted since the space limit.

4.4 Performance Evaluation

4.4.1 Simulation Setup

We consider a SSP based multi-hop CRN with multiple CR mesh routers deployed randomly in a $400 \times 400 \, \text{m}^2$ area. Suppose that the path loss factor $\alpha = 4$, the antenna related parameter $\beta = 4$, and the noise power at each router $\gamma = 10^{-9} \, \text{W}$. The transmitted power at each router is assumed to be equal as 10 W, i.e., $P_i = 10 \, \text{W}$, $\forall i \in \mathcal{N}$, and the transmission/interference range of each router are assumed to be 100 and 150 m, respectively. Furthermore, we assume that all available bands in the network have identical bandwidth as $W_m = 10 \, \text{MHz}$, $\forall m \in \mathcal{M}$, and the available band set of each router is set randomly as a subset of all available bands. Each SU has one session with a random rate demand within $[30, 90]$ Mbps and the source/destination nodes are chose randomly among the routers. The formulated MILP problems for resource allocation are solved by CPLEX [1].

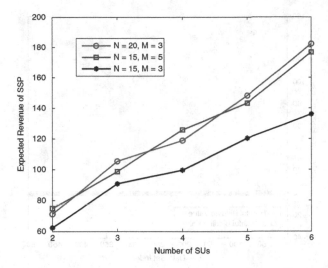

Fig. 4.1 Expected revenue of SSP versus the number of sessions with bidding for whole session

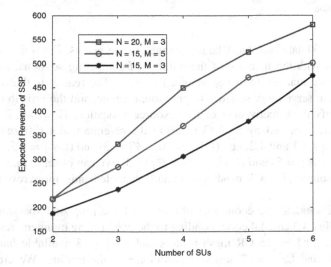

Fig. 4.2 Expected revenue of SSP versus the number of sessions with bidding for unit rate

4.4.2 Results and Analysis

For the two bidding manners, in Figs. 4.1 and 4.2, we show the expected revenue of SSP, i.e., the total bids of winners, versus the number of sessions (bidders) with different amounts of CR mesh routers ($|\mathcal{N}| = 15, 20$) and total available bands ($|\mathcal{M}| = 3, 5$) in the CRN. The bidding value of each session is randomly set within $[100, 150]$ for whole session in Fig. 4.1 and within $[3, 10]$ for unit rate in Fig. 4.2.

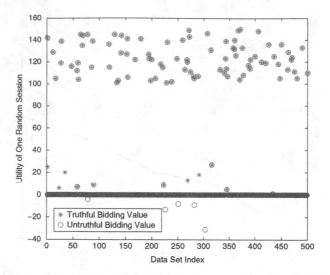

Fig. 4.3 500 data sets of utility of one random session with truthful and untruthful bidding value for whole session

We employ 100 data sets and take the average as the results. From the figures, we can find that with the increase of the number of competing sessions, the revenue of SSP in both manners will be enhanced as well. The reason is that for SSP, it always tries to serve more sessions to get more revenue, and the growth of revenue can just verify the effectiveness of the resource allocation. However, it is subject to the resources owned by SSP. Comparing the revenue under different network settings in Figs. 4.1 and 4.2, i.e., $(|\mathcal{N}| = 20, |\mathcal{M}| = 3)$ and $(|\mathcal{N}| = 15, |\mathcal{M}| = 3)$, $(|\mathcal{N}| = 15, |\mathcal{M}| = 5)$ and $(|\mathcal{N}| = 15, |\mathcal{M}| = 3)$, we can observe that when SSP has more resources, i.e., CR mesh routers and available bands, more revenue can be reached.

Next, we validate the economic-robustness of the proposed spectrum trading scheme in Figs. 4.3 and 4.4 corresponding to the two bidding manners, respectively. Assume that $|\mathcal{N}| = 15$ CR mesh routers and $|\mathcal{M}| = 3$ available bands are in the network, and $|\mathcal{L}| = 7$ sessions participate in the trading. We employ 500 data sets corresponding to 500 different network topologies. On each data set, we choose one session randomly and show its utility considering it bids truthfully and untruthfully, respectively. For the truthful bidding value, it is equal to the true valuation as a random value within [100, 150] in Fig. 4.3 and [3, 10] in Fig. 4.4. For the untruthful bidding value, it is equal to the true valuation adding a random value within [−100, 100] in Fig. 4.3 and [−3, 3] in Fig. 4.4. From both figures, we can find that when the sessions bid truthfully, their utilities are non-negative, which indicates that the trading is IR. Furthermore, when they bid untruthfully, their utilities cannot be improved, which means that the trading is IC and all sessions will bid according to their own true valuations.

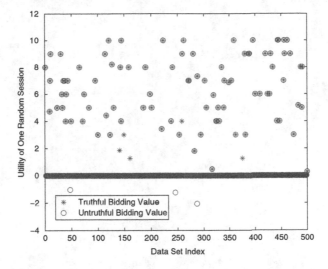

Fig. 4.4 500 data sets of utility of one random session with truthful and untruthful bidding value for unit rate

References

1. IBM ILOG CPLEX Optimizer.
2. P. Gupta and P. R. Kumar. The capacity of wireless networks. *IEEE Transactions on Information Theory*, 46(2):388–404, March 2000.
3. Y. T. Hou, Y. Shi, and H. D. Sherali. Spectrum sharing for multi-hop networking with cognitive radios. *IEEE Journal on Selected Areas in Communications*, 26(1):146–155, January 2008.
4. H. Li, Y. Cheng, C. Zhou, and P. Wan. Multi-dimensional conflict graph based computing for optimal capacity in MR-MC wireless networks. In *Proc. of International Conference on Distributed Computing Systems, ICDCS 2010*, Genoa, Italy, June 2010.
5. M. Li, P. Li, M. Pan, and J. Sun. Economic-robust transmission opportunity auction in multi-hop wireless networks. In *Proc. of IEEE Conference on Computer Communications, INFOCOM 2013*, Turin, Italy, April 2013.
6. M. Pan, C. Zhang, P. Li, and Y. Fang. Joint routing and scheduling for cognitive radio networks under uncertain spectrum supply. In *Proc. of IEEE Conference on Computer Communications, INFOCOM 2011*, Shanghai, China, April 2011.
7. J. Tang, S. Misra, and G. Xue. Joint spectrum allocation and scheduling for fair spectrum sharing in cognitive radio wireless networks. *Computer Networks (Elsevier) Journal*, 52(11):2148–2158, August 2008.
8. H. Zhai and Y. Fang. Impact of routing metrics on path capacity in multirate and multihop wireless ad hoc networks. In *Proc. of the IEEE International Conference on Network Protocols, ICNP 2006*, Santa Barbara, CA, November 2006.
9. X. Zhou, S. Gandhi, S. Suri, and H. Zheng. ebay in the sky: strategy-proof wireless spectrum auctions. In *Proc. of Mobile Computing and Networking, Mobicom '08*, San Francisco, CA, September 2008.
10. X. Zhou and H. Zheng. Trust: A general framework for truthful double spectrum auctions. In *Proc. of INFOCOM 2009*, Rio de Janeiro, Brazil, April 2009.

Printed in the United States
By Bookmasters

Printed in the United States
By Bookmasters